The Deep

Genre Fiction and Film Companions

Series Editor: Simon Bacon

THE

D
E
E
P

A Companion

Edited by Marko Teodorski and Simon Bacon

PETER LANG

Lausanne • Berlin • Bruxelles • Chennai • New York • Oxford

Bibliographic information published by the Deutsche Nationalbibliothek. The German National Library lists this publication in the German National Bibliography; detailed bibliographic data is available on the Internet at http://dnb.d-nb.de.

A catalogue record for this book is available from the British Library.

Library of Congress Cataloging-in-Publication Data

Names: Teodorski, Marko, editor. | Bacon, Simon, 1965- editor.
Title: The deep : a companion / [edited by] Marko Teodorski and Simon
 Bacon.
Description: Oxford ; New York : Peter Lang, [2023] | Series: Genre fiction
 and film companions, 2631-8725 ; vol no. 11 | Includes bibliographical
 references and index.
Identifiers: LCCN 2023015401 (print) | LCCN 2023015402 (ebook) | ISBN
 9781800792579 (paperback) | ISBN 9781800792586 (ebook) | ISBN
 9781800792593 (epub)
Subjects: LCSH: Sea monsters in literature. | Mermaids in literature. |
 Bodies of water in literature. | Ecocriticism in literature. | LCGFT:
 Literary criticism. | Essays.
Classification: LCC PN56.M55 D44 2023 (print) | LCC PN56.M55 (ebook) |
 DDC 809/.9336--dc23/eng/20230606
LC record available at https://lccn.loc.gov/2023015401
LC ebook record available at https://lccn.loc.gov/2023015402

Cover design by Brian Melville† for Peter Lang Group AG

ISSN 2631-8725
ISBN 978-1-80079-257-9 (print)
ISBN 978-1-80079-258-6 (ePDF)
ISBN 978-1-80079-259-3 (ePUB)
DOI 10.3726/b18081

© 2023 Peter Lang Group AG, Lausanne
Published by Peter Lang Ltd, Oxford, United Kingdom
info@peterlang.com – www.peterlang.com

Marko Teodorski and Simon Bacon have asserted
their right under the Copyright, Designs and Patents Act,
1988, to be identified as Editors of this Work.

This publication has been peer reviewed.

Contents

Contents

Contents

Acknowledgements

To begin we would like to say a huge well done to all the contributors to the book for actually getting their essays completed under such difficult conditions that the world has thrown at us over the past few years; it has been quite a journey since the idea for the book took shape and mermaids became merfolk, and they were joined by sea monsters and so much more. We would also like to thank Laurel at Peter Lang for all her help and patience during the production and completion of this book.

Marko: I would like to thank my colleague and co-editor Simon Bacon for his patience, perseverance and truly extraordinary intellectual and work stamina needed for this collection of essays to ever see the light of day.

Simon: Many thanks to my co-editor Marko who has been incredibly easy to work with through the entire process. I also want to thank the most important person in everything I do, my amazing wife Kasia, for her unending help, patience and support and without whom none of this would get done or be worth doing. Also, our two ever-growing monsters Seba and Majki who always manage to provide some light relief and distraction no matter how stressful things get. And last but by no means least the constant support (and sernik Magdy) of Mam I Tata Bronk.

Marko Teodorski and Simon Bacon

Introduction

In an age of accelerating climate change, rising sea levels and continents of garbage clogging the oceans, it is hardly surprising that creatures of *The Deep* are everywhere in twenty-first-century popular culture. From creature features like *Sharknado* (Ferrante, 2013) to underwater kingdoms and beings such as *Aquaman* (Wan, 2018), sea monsters and sea-people appear in fantasy series, horror films, children's cartoons and in the blanket ubiquity of Disney marketing for franchises such as *The Little Mermaid*. This is of course without mentioning the burgeoning industry of cosplay and costuming and the growing relevance of such figures in eco- and environmental messaging. Creatures and people from the sea are historically almost as old as humanity itself and, arguably, sea monsters and sea-folk are more globally popular and feature in more disparate cultural heritages than any other monster, including vampires.

In fact, this collection was born of the liminality of merfolk and their existence between worlds; human/non-human, land/sea, air/water. This makes them like "us" yet not "us" so they become points of anxiety and inquiry both in our relationship to the environment (Bacchilega and Brown 2019: xi–xii), and the nature of what it is to be human. What sea creatures, water-folk, and maybe more-so mermaids, also do is speak to sexuality and gender, both in terms of providing a queer non-human space of "possibility and radical imagination" (Sabrina Imler qtd in Braidwood 2023). But also the sexualising of the body of the Other reflecting from patriarchal cultures and their relationship to the wider environment: the exploitation of sea-brides having a direct correlation to abuses of Mother Nature. Consequently, merfolk and sea-people will feature strongly in this collection, though their connection to less explicit forms of anthropomorphism (such as water creatures and entities given human motivations and emotions) will be shown to example a similar mixing of the human and non-human – the meeting of species and "becoming" together – in relation to our environmental and sexualised relationship to *The Deep*.

This timely Companion will show just how widespread the belief in monstrous entities from *The Deep* is as well as putting them in historical and cultural context; it will highlight not only their ongoing importance both in terms of how we negotiate our own evolving sense of self in the twenty-first century but how we can reimagine our place alongside and entangled with other species and gain new insight to our fractured and broken relationship to the environment on which our future existence depends.

Environment, Context and Beginnings

The ocean covers 71 per cent of the Earth's surface and is the habitat of 230,000 known species, though as much of it is still unexplored, the total number of species has been estimated to go up to 2,000,000. Often considered the origin of life on our planet it has equally been seen as the home of sea monsters, sea-folk and beings from beyond our world – sometimes quite literally (see *The Abyss* (Cameron, 1989) and *Underwater* (Eubank, 2020)). Almost every culture around the globe treasures a story about a legendary, possibly godly or lethal, creature from *The Deep*. Maritime cultures generally prefer their monsters far away from the shore, but others find them in the depths of their rivers and lakes. The idea of *The Deep* is as culturally specific as it is transcultural so that while ships of the Scandinavian countries have been terrorised by the colossal squid Kraken, the inland countries dreamed about river nymphs, rusalki and lake or lagoon monsters that in equal measure enticed and seduced as much as they terrorised and tantalised.

It seems that the monster's size follows the size of its habitat, so it is natural that marine monsters come in sizes that mortal man can hardly cope with, if at all. For ancient cultures that imagined the Ocean as the vast Beyond, as the fluid matter that encircled the known, or possible, world, creatures of *The Deep* were also the creatures of dimensions utterly divested from their mortal existence. One of the most common motives in maritime cultures is, thus, that of *Chaoskampf*, depicting a battle of a hero deity with a chaos monster, often in the shape of a serpent or dragon. The Sumerian Sea serpent Lotan dies at the hands of the storm god Hadad-Ba'al; Leviathan is slain by Yahweh in the Hebrew Bible; primordial Babylonian goddess of the salt sea Tiamat perishes in the battle with Marduk; Zeus kills Typhon; Thor battles the World Serpent Jörmungandr; Indra kills the Hindu monster Vrtra; Slavic peoples to this day celebrate St George and his battle with the Dragon – originally a lake serpent. In all the examples, sea monsters signify the primordial chaos as the ultimate challenge to the world order. They are not only embodiments of culturally specific narratives and values but also of the transcultural struggle of men with unyielding, unknown and unpredictable cosmic forces of nature, life and death.

Sea creatures have plagued the Western imagination since its presumed classical roots (and before), but in the Middle Ages their role as signifiers of the unknowable became part of a historical and colonial narrative. On navigation maps, unexplored places of the ocean were populated with images of their horrific bodies (like 1539 Carta Marina, for instance), followed by signs "here be monsters." Numerous sightings and encounters embedded them deeply into the cultural imaginarium, the pool of images and metaphors for the expression of the terror of the unknown deep, so much that today we seem incapable of imagining *The Deep* without them.

However, there is a group of underwater beings whose relationship to humans is different from the rest of the dwellers of *The Deep*. While sea serpents, giant squids and dragons are godly, theogonic and metaphysical, seafolk tend to be of a size more easily understood by mortals. Merfolk – sirens, mermaids, mermen, rusalki, nymphs, selkies, tritons – are creatures that also inhabit the outer rim of identity, of the social and of the acceptable, but possess an extraordinary ability to change. So, sirens and mermaids – merfolk or seafolk in general – changed radically through the history of the West, as well as throughout the world. In the pre-classical and classical times, sirens were omniscient bird-women, daughters of the dark branches of the Greek theogonic tree; they lured men by dulcet voices and let them perish at the edges of their rocky island. In the Middle Ages, they acquired tails, began feeding on men's souls and merged with mermaids, to this day confusing readers as to the difference between the two. More so, they disturbingly blurred the differences between human and non-human: they were dangerous as they disturbed the rigid categorisation of the world where mankind stood distinct and at its pinnacle while simultaneously offering a glimpse of otherness that lived beyond the rules and restrictions of society. Water-folk, whether in rivers, lakes or the ocean, have understandably captured the imagination of nearby communities and even entire nations as they acted as a bridge between humanity and the creatures of *The Deep*, as well as something of a gauge of our relationship to the watery environment they come from and our level of dependence upon it.

Imagining Merfolk

Examining their history in its entirety is beyond possible, considering that merfolk span the whole history of humankind, defying the natural borders created by oceans, mountains, rivers and continents. There are no imaginary and mythical creatures more familiar to people in all corners of the world, in almost every known historical era (see Austern and Naroditskaya 2006); the Syrian goddess Atergatis was known as a "fish-goddess," and the Babylonian Ea or Oanness was represented as part man and part fish (Waugh 1960). Merfolk, in their local versions, roam the Andes as well as Russian lakes; *ningyo* (human fish) are known in Japan, fishtailed Mami Wata spirits are worshipped throughout the South America and African diaspora, while mermaids are spotted combing their hair and beckoning to sailors in both northern and southern seas. And in most accounts, their beauty and voices are pervasive, the ecstasy they offer is unending: existentially unbearable and historically indestructible. And as if this geographical and historical omnipresence were not enough, merfolk assumed another unsurpassable feature: a staggering ability to change. Ancient, feathered enchantresses or medieval fishtailed whores, merfolk never cease to morph, mutate, to transcend their impossible corporeal existence, merging into one another.

The nineteenth century played a crucial part in the merfolk's contemporary omnipresence. The "mermaid craze" began with P. T. Barnum's Fiji Mermaid 1822 hoax (a whole little industry of these grotesquely appealing products existed in Japan, supplying the British market) (see Carrington 1957;

Cook 2005), but it was Hans Christian Andersen's *Little Mermaid* that immortalised them in the Western imagination. Although its versions had existed in Central and East European literature and opera (see the Rusalki chapter of this collection), only with the 1872 English translation the Anglophone audience was introduced to the loving, compassionate being we are familiar with today, a complete opposite to its species three millennia-long history. And after its twentieth-century Disneyfication, sirens and mermaids are literally everywhere. People dress up as merfolk for the iconic Coney Island Mermaid Parade (Coney Island 2022; Hayward and Milner 2018); in major European and American cities at every corner there is a Starbucks coffee shop with its two-tail mermaid shining bright.

At the beginning of the twenty-first century, the mermaid (merman, merfolk) as a creature and as an idea, has become a profession, immersed deeply into the fabric of consumption and capitalist desire (see Sax 2000). There is a plethora of "professional mermaids," (Melisa 2022) mostly underwater performers and custom fishtail designers creating an industry where large amounts of money can be made (some Eric Ducharme's custom-designed tails cost $2,759 and his customer list includes celebrities, like Lady Gaga; Tungol 2013). Similarly, professional mermaids and merman are promoted as icons of beauty, voluptuousness, sexuality and allure, their image and their bodies selling the products they advertise or rallying people around an environmental cause. Their monstrosity is erased: little girls want to be mermaids. They advertise designer shoes despite the fact that they do not have legs; their bodies are turned human, appropriated as signs referring to consumption. In the 2020s being a mermaid is a valid career choice.

However numerous in contemporary culture, merfolk are only a fragment of the (post)modern audience's infatuation with aquatic monsters. From cinema, through comics and literature, to everyday commodities (such as the mentioned coffee mugs), water remained the inexhaustible pool of fantasies for the modern consumer. The superhero Aquaman alone spans eighty years of comics, animation and feature films, culminating in the 2018 Warner Bros' blockbuster and its much-awaited sequel (slated for 2023), and Abe Sapien (the "amphibious man") spawned from the *Hellboy* franchise in the 1990s and since starring in a comic of his own. Less contemporary manifestations equally caught the imagination of the times that created them, such as Gill-man, who

Introduction

was extremely popular in the 1950s, so much so that *Creature of the Black Lagoon* required two sequels, and numerous cameo appearances (including an episode of *The Munsters* (Burns, 1964–6), the motion picture *The Monster Squad* (Dekker 1987)) – more recently his daughter Lagoona Blue features in the hugely popular children's series *Monster High* (2010–17).

There is an important issue at work here. Sirens have been luring men to their doom since the time immemorial, so in the era of high capitalism it seems only natural that their powers have been transformed into visual pleasures of commodities. This has been happening over the entire twentieth century. But monstrosity became a central trope of the contemporary entertainment industry and that requires an explanation. If we start from the premise that the infatuation with *The Deep* reflects our personal desire for the undiscovered, hidden, or repressed part of the self, what does this mean for the contemporary consumer who awards the Oscar for Best Motion Picture of the Year to Guillermo del Toro's *The Shape of Water* (2017), a story of an amphibious, humanoid creature trapped in the human world? What does this new fashion of people spending money on custom-tailored tails and dressing up as mermaids means for the relationship of the contemporary consumer to her/his unconscious?

Of course, beneath the surface of commercial success, a more complicated and sometimes disturbing creature emerges from *The Deep*, and one often fuelled by desire, though also revenge. Desire takes the form of transgressive sexualities and couplings, with the aforementioned *The Shape of Water* being a good example of this speaking both to various kinds of otherness – the muteness of the female lead Eliza (Sally Hawkins) and the Amphibian Man (Doug Jones) – and a sexual relationship that is literally "out-of-this-world." Female sexuality and sexual pleasure have long been represented through the interaction between the female body and the ocean, partly through metaphorical correlation, but also the pleasures provided by, and embodied in, multi-limbed cephalopods and hard-underwater male bodies – *The Shape of Water* spawned a range of female sex-toys (see Sharf 2018). The sexual prowess of the subalterned body signifies both the monstrosity of the acts performed with it and the anxiety inspired in white, patriarchal minds (see Slatton and Spates 2014). But sexuality looms large in the popular imagination of sea "monsters" and water-people equally as a means of containment and agency, with literary mermaids often forced to conform to human norms to remain on land, whilst others finding it a means to go beyond societal norms and expectations – the sea itself being equated to a non-hospitable, non-human space, but one where the transformative, the exotic and the queer live. Masculinity also finds expression through the monstrous marine body as hapless victim or ancient god, though one that often more clearly reveals the otherness of humanity itself through mimicking and mirroring the mores of human society in the body of the other.

Of course, the mixing and meeting of species is central to much of our relations to *The Deep* and the creatures, real and imagined, that populate it, which has consequences for further theoretical considerations and interpretations. One such is Critical Animal Studies (CAS), though this is a linkage that is not without its complications. CAS, as noted by theorists such as Dawne McCance (2013), John Sorenson (2014) and Linda Kalof, is constructed around an opposition to speciesism and the promoting of animal rites along the lines of ethical and judicial equality. However, much of our historical representation of creatures from *The Deep*, and more so hybrid ones, is one of distinct inequality and othering. The meeting of species "monsterises" both, a positive in monster theory, but considerably less so in terms of animal rights and the recognition of other species as being of equal worth to humans. For Andersen's Little Mermaid, her quest for the Prince is as much about gaining a soul so that she might become human – the excruciating pain in exchange for losing her connection to the ocean is one she accepts willingly.

In a similar vein, creature features of the 1970s onwards represent water life such as sharks, crocodiles and piranhas as a source of danger to humanity, and though this can be read through the lens of eco-revenge and the ecogothic, it still paints nature in general and aquatic life in particular as antithetical to humans. However, one might take the thread of otherness implicit in this, where mankind is just another food source, and extrapolate to the idea of humans are just another animal, hence reinforcing a sense of difference that purposely rejects easy forms of anthropomorphism. Contrarily one can also read certain kinds of anthropomorphism and interspecies empathy as a way to provide a measure of respect, however misplaced that might be.

Arguably, of more relevance here is the related topic of interspecies hybridity, and Donna Haraway's ideas around the meeting of species. Donna Haraway cites the multitudinous nature of humans and, indeed, all species, which are never singular but are in fact legion. Here, our bodies are not only constituted of many other organisms, microbes and lifeforms but are historically and culturally situated as well. Consequently, we are internally and externally entangled entities that then encounter other internally and externally entangled entities (like Haraway's friend's dog; 2012: 6–8). Further to this, once we consider these entities in their constituent parts, we are all far more similar than we are different – humans have the same number of genes as do sea urchins – revealing our embeddedness within the natural world around us and the ties that bind us to others and the animal Other (Govindrahan 2018: 3–4).

Rosi Braidotti continues in similar vein, though adding a Deleuzian inflexion where the animal and the non-human are an evolutionary imperative towards the post-human and the becoming-animal – one could add becoming-nature to this as the (possible) human is "fully immersed in and immanent to a network of non-human (vegetable, animal, viral) relations" (2013: 193). All of these critics see the living and the alive that are not human as companion species (Haraway 2012) or environmental companions. Here then, in this vision of the post-human, our similarities unify us, so that the mermaid is no longer an abomination of nature and mankind's dominion over the world, but both symbolic of our sharedness (biological equality) and entanglement, and a literal representation of our future selves.

The idea of critique mentioned above points to the forms of queering and questioning that are fundamentally part of the water-creatures essence as it is of all monsters, whose etymologically designed purpose – *monstrum* meaning portent/sign – is to reveal. The revelation provided by water-folk and sea monsters in an age of often unrestrained consumption and waste, is how deadly we have made their environment, which is a harm not just visited upon them but also ourselves. In this sense, there is much in the body of merfolk that intimates a return to the past, a mysterious earlier "normal" where the human and non-human worlds were more evenly balanced – a time before the consumerist society mentioned above, even though it is that same consumerism, via "mermaiding," that provides the means to enact such performative returns (Launier 2010). Environment, on the other hand, is often imbued with human emotions, primed to defend itself at all costs (see Soles and Tidwell 2021), and takes the form of malicious or mutated sea creatures, monsters from *The Deep*, or unknown races of sea-people claiming their rightful place at the top of the food chain within (or against) the Anthropocene/Capitalocene. However, what is truly remarkable are the ways in which the sea and its inhabitants can be representational of a new age and a more symbiotic future with humanity. While much of this is currently limited to fiction, aquatic diseases and parasites regularly cross the water/land divide infecting and mutating their hosts and sudden changes in the environment of the earth, such as rising sea levels, could see humans having to evolve to life in, and under, water (Bates 2011).[1] This oddly harks back to earlier tales of Selkies and even Melusine where the seal-wife or mermaid, respectively, has human children with the man who has captured her. However, in the post-human version of this "fairytale for the future" the children would be neither one thing nor the other, neither flesh nor fish, but a new creation: a step in an ongoing evolution.

In the 2000s the contemporary immersion with the "world of the sea monster" compared to nineteenth-century cautionary tales seems ridiculously naïve. We do not need to read the monster "paranoically" anymore. It is not hidden within the language, in the cracks and crevices of representation; it does not hide from view. If psychoanalysis opened up the unconscious and presented it to the Western subject, the entertainment industry, as well as the arts, has furthered this process to an extreme. As the monsters of *The Deep* live more and more in the open of our everyday lives, so our inner, forbidden desires rise with them, and as importantly the possibilities, freedoms and transformations they offer to emerge from the depths and be who we naturally are and who we might become; the monster is no longer within, but an inherent part of our existence. In the twenty-first century the sea-folk and sea monsters of *The Deep* no longer describe the alienation and repression of the self or societal excommunication, but rather become a symbol of individuality and agency. In this sense, sea-folk in particular are not just a signifier of where we have come from, but an indicator of where we might be going.

---

[1]   There have been theories that early human development might have included divergents that were more suited to aquatic life (see Rhys-Evans 2019).

Introduction

Frameworks of Engagement

The Companion will chart our ongoing interactions with creatures from *The Deep*, be they merfolk or sea monsters, with particular focus on the environments from which they come (cultural and historical) and also those they symbolise (the lake, or sea in particular, and nature or wider ecology in general). Ideas around gender and sexuality play into that and are inextricably linked to other areas of individual and collective identity such as ethnicity, disability, or otherness. The Companion will begin with historical and national/regional precedents and will progress in a chronological manner – with the occasional aberration as befitting the aberrational nature of seafolk – up to the twenty-first century and intimations of what is to come. This mapping will consist of thirty-one original essays that take a single example/text/film/artwork as a lens to look at the wider topic. These are divided into six parts, though many can be seen to link through and resonate with papers from other sections. Indeed, there is certain inter-connectivity between the creatures of *The Deep* and our relationship to them over time that makes distinct groupings problematic, though hopefully the ones chosen here will act as a starting point with which to explore further.

The collection begins with "Part 1: Cultural and Environmental Beginnings" which looks at six different global locations and their own particular manifestations of sea monsters or merfolk and the cultural environment from which they emerged. The first essay, "Biblical Myth: Leviathan, Mythical Creature (600 BC–present)," by Brandon Grafius looks at the possibly oldest of the huge sea monsters, Leviathan. Coming from the creation myths of many cultures the author traces its journey through the Hebrew Bible as signifier of cosmic chaos, up to the present day where it configures an otherness from beyond our world. This is followed by "Indian Myth: The Matsya Avatar (300 AD–present)," by Debaditya Mukhopadhyay that features one of the earlier manifestations of a "merman" as an avatar of the god Vishnu. With a curious resonance to biblical myth, Vishnu in a half-man/half-fish form guides the remnant of creation to safety during a great flood. Next, Manal Shalaby in "Middle Eastern Myth: *Scales* (Shahad Ameen, 2019)," considers a recent Saudi film and how it utilises Middle Eastern myths around female deities and the sea. With their links to human sacrifice, the author further relates how ancient ritual still informs contemporary environments. Moving to the Americas, Amylou Ahava's, "Iroquois Myth: The Great Horned Serpent (c. 1450–present)," examines an indigenous myth that mirrors the wider manifestation of giant serpents or snakes in global creation myths. Here the Great Lakes of North America provide a distinct geographical character to the monster that hides below the surface. Next is Marko Teodorski and "Slavic Myth: *Rusalki* (Witold Pruszkowski, 1877)," which looks at the freshwater creatures the rusalki. In the Slavic tradition they are not hybrid entities but rather romanticised figures of women who have drowned, and are represented through a peculiar semantic network of paleness, moonlight, sleep/reverie and stillness. The first part closes with "Modern Japanese Myth: *Tropic of the Sea* (Satoshi Kon, 2013)," by Leila Anani who focuses on the appearance of merpeople in Japanese Manga and Anime. More specifically the author picks out environmental concerns as a key feature of these texts and highlights how they utilise features from Japan's own myths around fish-people from the sea.

The next section, "Part II: Femininities and The Deep," focuses on issues around the mermaids and femininity in relation to the patriarchal societies that have tried to contain them and also in ways that they have found agency within that. This begins with a rather traditional view of the connection between the mermaid and the ocean as seen in "Aquatic Femininity: *Duyung* (Abdul Razak Mohaideen, 2008)," by Philip Hayward that examines Malaysian myth around fish-peoples of the various coasts and island of the nation and their recent representation on film. Further, the author describes how regional variations of the myth remain resistant to more global interpretations. Following this is "Punishing the Monstrous Feminine: The Legend of Kópakonan (1891)," by Laura Sedgwick begins with a legend of a seal bride from the Faroe Islands: a tale of a sea-woman kept on land against her will and forced to have children, her sea-family threatened by the man that holds her captive. Yet she is to blame when she finally escapes. In "Feminine Magics: *The Little Mermaid* (Hans Christian Andersen, 1837)," Daisy Butcher turns to one of the most famous mermaid texts, though does not focus on the nameless sea princess, but the villainised or "hagsloited" sea-witch. However, as the author argues, her villainy is not in her "witchiness," but rather in her free expression of sexuality and personal pleasure. This is followed by Martine Mussies' "Reclaiming the

Feminine: #Posidaeja (Efa, 2021)," that considers the ways in which female fan art of mermaids has increasingly picked up and utilised references to the earliest myths involving mermaids and goddesses connected to the sea, in particular the Greco-Roman goddess Posidaeja. "Part II" closes with "Feminine Self in the Male Psyche: *Underwater* (William Eubank, 2020)," by Phil Fitzsimmons which describes a journey into death and the monstrous feminine as seen in the film *Underwater*. This further sees the female lead find both herself and obliteration in the decision between the all-consuming feminine and the male psyche.

"Part III: Masculinities and The Deep" more closely considers the ways in which sea-folk, mermen and Gil-men represent varying forms of masculinity. The section begins with "Changing Masculinities: *Aquaman Volume 1: The Trench* (Geoff Johns, Ivan Reis and Joe Prado, 2011–2012)" by Carl Wilson, who studies the figure of Aquaman, and how the kinds of masculinity he represents have changed over time on his appearances in Marvel Comics and onto the cinema screen. This is followed by Gerard Gibson's "Contested Masculinities: *The Creature from the Black Lagoon* (Jack Arnold, 1954)," examination of, arguably, the most famous iteration of the Gil-man. The Creature is shown to have a very particular relationship to the times that created it and the environment it lives in. "Recontextualising the Past: *The Shadow Over Innsmouth* (H. P. Lovecraft, 1936)," by Brigid Cherry focuses on some recent adaptations of Lovecraft's story of underwater creatures from beyond our world and how they address the very specific construction of white manhood envisioned in the original. "Part III" ends with "Queer Affect: Hans Christian Andersen (1805–1875)," by Kodi Maier that considers how the author of *The Little Mermaid* expressed his own sexuality through the figure of the underwater princess and how this same "queering" has followed the story into its later Disneyfication.

The next section "Part IV: Identities and Difference," considers the ways in which the otherness of sea creatures and sea-folk allow for the creation of alternative identity positions as fictional example and real-world situation, both in terms that recalibrate our wider relationship to *The Deep*. "Mimicking Femininity: *Into the Drowning Deep* (Mira Grant, 2017)", by Agnieszka Kotwasińska looks at the recent novel by Grant that subverts the expected tropes of the mermaid story, revealing the sea-wives to actually be male predators revealing the performative nature of gender. "Agency and the Feminine Body: *The Little Mermaid* (Hans Christian Andersen, 1837)," by Astrid Crosland re-reads Andersen's classic tale as one of personal choice and individual transcendence. This is followed by "Environmental Disability and Aquatic Climate Crises: 'The Mermaid' (Hanna Cormick, 2018)," by Alison Sperling, that looks at the work of performance artist Hanna Cormick who uses the physical limitations and non-normative aspects of herself in relation to the otherness of the mermaid body to investigate and comment on our own precarity and that of the environment around us. Ultimately Cormick calls for recognition of our complete reliance on environment and that caring for it is caring for ourselves. Alison Patterson's "Disentangling Difference: *Song of the Sea* (Tomm Moore, 2014)," examines Irish myth and the story of the child of a selkie – unusual in that such narratives more often focus on the seal-wife rather than her part-human children. Here difference and otherness become a means of experiencing life and the environment beyond the strictures of normative society. "Beyond Human Desire: *Possession* (1981) to *My Octopus Teacher* (2020)," by Jon Hackett brings "Part IV" to a close with his study on a recent and popular documentary on one man's relationship with an octopus. Spectacle and the gaze, at least from the human standpoint, are central here and simultaneously point to earlier forms of the sexualisation of cephalopods, but also ways in which our approach to the environment might be reconfigured.

"Part V: Human Incursions and Environmental Responses," looks more specifically at instances of what might be termed eco-revenge due to human incursion into the "home" of the sea monster. It is not coincidental that such territorial invasions are often configured around colonialism and exploitation. This part opens with "Oceanic Epistemologies: The *Daedalus* and the Great Sea Serpent (1848)," by Jimmy Packham, that looks at the most well-known nineteenth-century sighting of sea serpents which coincided with the laying of telegraph cables across the ocean floor – an act that simultaneously exampled an attempt at containing the unknown while revealing how impossible that task is. Next is "The Eco-other as Spectacle: *The Lure* [Córki dancingu] (Agnieszka Smoczyńska, 2015)." Lauren A. Mitchell focuses on a different facet of otherness, in its uses and abuses of the idea of spectacle. Specifically, how the otherness of the aquatic body, both visually and sexually, is what provides it with visibility and if that otherness is physically removed, it is no longer "seen" and loses its ontological purpose and meaning. This is followed by Matt Melia's "Environmental Exploitation: *The Meg* (Jon Turtletaub, 2018)" that similarly examines human exploitation of the ocean, but in this case the vastly underexplored Mariana Trench – which acts as a point of concentration of the dangers of *The Deep* in many texts included in this collection. From here the eponymous Meg (Meglodon), prehistoric shark emerges as ecological revenge and ideological mirror to the capitalist "shark" that facilitated its release, consuming everything in its path. "Neo-colonialism and the Liminal: *Into the Drowning Deep* (Mira Grant, 2017)," by Jennifer K. Cox takes a differing perspective on Grant's novel reading the emergence of the aquatic threat, also from the Mariana Trench, as one of colonial resistance and mimicry. Consequently, the monsters from *The Deep* perform an act of reverse-consumerism "eating" those that would consume their world. Next Catherine Pugh's "Ecological Decolonisation: *Crawl* (Alexandre Aja, 2019)," considers a more dispassionate, less anthropomorphised, vision of ecological revenge where nature and its water creatures just simply moves back into areas that were formerly theirs. Humans here are not shown as the enemy of nature, but rather a snack to be enjoyed as the alligators move back into their old family home. Tom Ue in "Ecological Exchanges: *Spirited Away* (Hayao Miyazaki, 2001)," considers more closely the interconnectedness between us, our environment and the natural world, as seen in the river spirit Chihiro, showing how they co-exist through a series of economical exchanges. Mayazaki's film suggests that the memory of our previous ecological relationships offer some form of hope within that, yet not enough to bring us back into balance.

The final section, "Part VI: Ecological Entanglements and Environmental Futures," looks more closely at environmental and ecological issues around the ocean suggesting positives and negatives within that and also ways in which our entangled existences can configure different approaches to our shared futures. "Ecofascism and the New World: Abe Sapien (1994–present)," by Tom Shapira looks at the popular, recent, iteration of the Gil-man and his individualistic approach to hybridity and difference and what that says about possible eco-futures. This is followed by "The Healing Ocean: *Moana* (Ron Clements and John Musker, 2016)," by Kevin J. Wetmore, Jr. who considers the Disney adaptation of a mixture of Polynesian myths in the story of Moana. Here the sea is highlighted as essential to human life as well the repository of essential ecological knowledge. Ruth Barratt-Peacock, in "Becoming Ocean, Becoming Self: Ocean Poems (David Malouf, 1976– 1991)," takes this further in a close reading of some of the poetry of Australian poet David Malouf. Here the poet suggests that our ongoing relationship and entanglement with *The Deep* is one beyond the languages that we currently know or understand. "Transgressive Reproduction: *Évolution* (Lucile Hadžihalilović, 2015)" by Octavia Cade purposely pushes back on human exploitation of the oceans where a mysterious form of mermaid, or humanoid sea creature, utilises male human children as incubators for their own offspring. The film has no answers to what this might mean for the future or whether it is one that humanity might want. The collection closes with Justin Wigard's "Envirofuturism: *Subnautica* (Unknown Worlds Entertainment, 2014)," which reveals the dilemma of humanity and the oceans in the game space of a distant planet. Survival can only occur by understanding and working with the ecological balance of the largely aquatic terrain and more so ensuring that the planetary matriarch, the Sea Emperor Leviathan, remains alive. Thus, completing the circle that began in the first essay featuring the biblical monster Leviathan, a patriarchal monster that threatened the survival of mankind, and now closing with a matriarchal colossus that humanity cannot survive without.

This brings to a close this critical survey of sea-people and sea-monsters that lays out their historical and multicultural beginnings, as well as providing readings of their possible meanings and cultural relevance in the twenty-first century. More so, it details our fascination with *The Deep* as source and receptacle of human desire, ecological womb, eternal mother, and as an environmental *vagina dentata* that will bite off and chew the unwanted intrusion of patriarchal, imperial colonialism and exploitation. While the Companion contains some darker notes in our ever-evolving relationship to *The Deep*, it finishes on points of potential and hope where humanity can recognise its affinities with our watery "monsters" and our mutual reliance upon the environment. The timeliness of such a message cannot be overemphasised, but it also suggests that the continuing popularity of merfolk, sea monsters and mutated sharks is a message we already know if we are only prepared to read

Introduction

the signs. This Companion is then part of this process, in bringing together an extensive collection of historical and cultural sources whilst intimating further and necessary areas of ongoing and emerging study. *The Deep* is not just a record of our past, but a map to our possible future.

Image Intervention I:
Myths of the Sea and the Sky

Figure 1. Artwork by Derek Newman-Stille (Reproduced with permission).

Part I

Mythical Imaginings

Brandon R. Grafius

Leviathan, Mythical Creature (600 BC–present)

Of all of the monsters of the deep, the oldest might be Leviathan. This grand sea monster – depicted in various stories as a giant whale, a dragon, or a seven-headed hydra – is found in the creation myths of many cultures, from India to Iran to Mesopotamia (Miller II 2018). The monster is known under a variety of names, but a constellation of characteristics marks it as the same beast: it is the ruler of the sea and serves as the heroic god's most-feared antagonist. In the earliest myths, the monster is female, and is frequently depicted as the embodiment of chaos that the creator God must defeat to create the world. In the East Semitic creation myth known as the *Enuma Elish*, Marduk defeats watery chaos (here named Tiamat) and shapes the world out of her corpse.[1] In the Ugaritic myth, the monster is named *litanu*, a clear cognate to the Hebrew *levytan* (Korpel and de Moor 2017: 6–9). No matter the name, the linguistic connections between all of these myths make clear that this is a series of variations on the same primordial chaos monster (Ballentine 2015: 76–90; Marzouk 2015: 71–8). Leviathan does not appear directly in the Bible's creation narrative, but the monster swims around in the background. There is no direct conflict with Leviathan in Genesis 1, but the narrative still presents a struggle between order and chaos, similar to the battle undertaken against Tiamat or *litanu*. "In the beginning ..." the text says, "the earth was a formless void and darkness covered the face of the deep" (Genesis 1:1–2).[2] This seems to imply that creation did not occur out of nothing; before the order of the world

1 The oldest texts we have of the *Enuma Elish* are dated from the thirteenth to eleventh centuries BCE, though scholars have placed the date of original composition as early as the twentieth century BCE. See Batto (1992: 33–9) for a brief overview.
2 All biblical translations are from the *New Revised Standard Version*.

there was chaos (Keller 2003). Instead, creation is conceived of as shaping order out of the primeval chaos, represented as the sea. God overcomes the chaos by shaping it into order.

But Leviathan also makes more direct appearances throughout the Hebrew Bible, first on the fifth day of Genesis 1's creation story, when God creates the "great sea monsters" as the first named creature (Genesis 1:21; Mobley 2012: 16–33). There are references to God's battle with Leviathan scattered throughout the Bible (e.g. in Psalm 74:13–14; Isaiah 27:1; and also in Isaiah 30:7 and 51:9, where he is called Rahab; see Dekker 2017: 21–39), but here in Genesis Leviathan is simply a part of God's creation (Levenson 1988: 54–65). Several passages in the Hebrew Bible refer to Leviathan simply as "the dragon" (Job 7:12; Psalm 74:13; Isaiah 27:1, 51:9; Ezekiel 29:3, 32:2), which is the name used for his appearance in the New Testament Book of Revelation. Here, Leviathan is the enemy God will face in the eschatological war (Macumber 2019, 2021). So interestingly, Leviathan is the enemy God faces at the beginning, during the creation (in passages such as Psalm 74 or Isaiah 51), but also the enemy God must face in the final apocalyptic battle of the eschaton (as depicted in Isaiah 27 and Revelation). Leviathan seems like the earliest example of Jeffrey Cohen's famous dictate that "the monster always escapes" (Cohen 1996: 4–6), as it is envisioned as both the alpha and the omega of divine combat myths in these various texts. Leviathan is God's once and future opponent. But for Leviathan's continued importance for popular culture, no passage is as influential as Job 41.[3]

Most of the book of Job is a wisdom dispute between Job and his friends regarding the nature of suffering, the justice of the universe, and the presence of God (Newsom 2003: 72–89). But at the climax of the book, God speaks to Job out of a whirlwind, giving Job a tour of the wonders of the universe (Brown 2014: 109–27; Doak 2014: 183–232). The culmination of this extended speech is a description of Leviathan, consuming a full chapter. Leviathan is monstrous and threatening, yes; but also "awesome" (Ansell 2017: 90–117; Grafius 2019: 41–6). For God, Leviathan is a part of creation and a part that God is particularly proud of. As part of the speech, God says, "I will not keep silence concerning its limbs, or its mighty strength, or its splendid frame" (Job

3   The Hebrew versification is slightly different; for the sake of simplicity, I will refer to English verse numbers throughout.

41:12). God understands that Leviathan is a threat to the order of creation. In this tale, God has responded to this threat not by destroying Leviathan, as in some of the monster's other mythological appearances, but by bounding Leviathan within the depths of the sea. But in addition to being horrifying, Leviathan is also deeply awesome. Leviathan then can also be seen as sublime, in all its overwhelming terror, in all its awe-inspiring grandeur.

The concept of the sublime is usually traced back to Immanuel Kant, whose writing is then built upon by Edmund Burke. The concept most clearly enters the discourse of religion through the work of Rudolf Otto, in his influential book *The Idea of the Holy* (1917).[4] For Otto, much of theology had become concerned with the rational side of God, the side that could be readily understood through human concepts. Otto wanted to focus on the non-rational side of God, the side he termed the "numinous" – and which contains experiences that can be seen to be akin to that of the sublime. It is this side which leaves us overwhelmed and filled with awe, wonder and terror, since we cannot necessarily trust the benevolence of this God's universe. Indeed, Otto suggests that encounters with the numinous leave us deeply aware that our existence depends upon the continued permission of this inscrutable God. In the Hebrew Bible, one of the clearest reminders of this God's ultimate otherness comes in the form of Leviathan, the living embodiment of chaos that God nevertheless grants a place within creation. In the order of the world, even ultimate chaos has a place. Sean Moreland has argued for a distinction between the Kantian/Burkean sublime and Lovecraftian "cosmic horror" which Moreland traces back to Lucretius (Moreland 2018: 13–42). In Moreland's reading, the sublime includes both awe and terror, but eventually the feeling of awe overwhelms that of terror, and the experience is resolved into that of pure awe. But in cosmic terror, these two emotions – awe and terror – continually exist in a paradoxical relationship, neither overcoming nor outweighing the other. In the biblical texts, and many myths of the Ancient Near East, the figure of Leviathan is sometimes a figure of cosmic terror, and sometimes a sublime figure of marvel. But however, the text resolves these emotions, the monster is accompanied by a swirling blend of overwhelming awe and dread. In some of the biblical texts, such as Genesis 1 and Psalm 104, Leviathan is a figure of the

---

4   The first English translation appeared in 1923.

sublime. In Job 41, I would suggest that God views him as a sublime creature, while suggesting that he is a figure of cosmic terror for the rest of creation.

It is this connection with cosmic terror, the overlap of horror and grandeur, that most clearly marks Leviathan's return to the depths in contemporary cinema. Reinier Sonneveld has written on Leviathan in film, but his definition is so broad that he scoops up figures as diverse as the Death Star from the *Star Wars* franchise and the swirling hordes of great white sharks of *Sharknado* (2013) in his net (Sonneveld 2017: 280–95). It's difficult to imagine a film as self-consciously goofy as *Sharknado* having anything to do with the sublime or cosmic terror. For Sonneveld, any "incarnation of death" can be read as a Leviathan figure. I would prefer, instead, to narrow the focus as follows: Leviathan appears in film and popular culture as a deep-sea monster, which does not bother humanity if left alone, but responds with incredible violence to humans wandering into its domain. Furthermore, this creature must instil both fear and awe; not only terror at the prospect of physical death but also a reconsideration of humanity's place within the order of the cosmos. As Otto describes the numinous, it is the feeling of "impotence and general nothingness as against overpowering might, dust and ashes as against 'majesty'" (Otto 1950: 21). In contemporary film Leviathan is more a figure

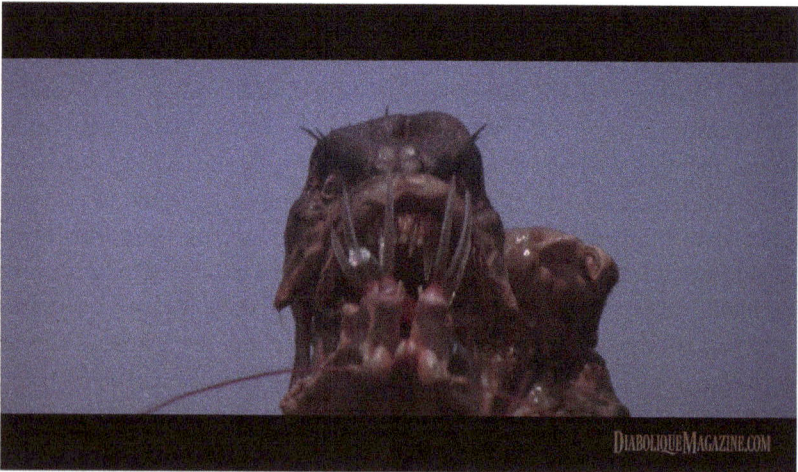

Figure 2. Rubber-suit sublime in *Leviathan*, directed by George P. Cosmatos (Metro-Goldwyn-Meyer, 1989).

of cosmic terror than of the sublime, the majesty against which humanity has no hope. Our only chance is to stick to our own spheres of existence and leave Leviathan alone in the ocean.

Science fiction and horror films have long imagined horrifying creatures emerging from the unexplored depths of the sea, such as the giant squid of *20,000 Leagues Under the Sea* (1954). The late eighties saw a boom in this sub-genre, with *Deep Star Six* (1989) and *The Rift* (1990) both featuring prehistoric, predatory sea creatures. There was even a film called *Leviathan* (1989) (see Figure 2), but it has more in common with John Carpenter's *The Thing* (1982) than with a cosmic struggle between order and chaos. While some scholars (Hauser 2018) have placed *The Thing* in the broader category of "weird cinema," the creature in Carpenter's film does not approach the awe-full, overwhelming numinous of which Otto speaks.

Of course, the category of the cosmic horror is somewhat in the eye of the beholder. But I would argue that most of these films remain firmly planted in the realm of mortal terror, with the primary focus being on the life-or-death struggles of the protagonists. The monsters may (or may not) be horrifying, but the horror is about what they can do to human bodies. A giant crab (as

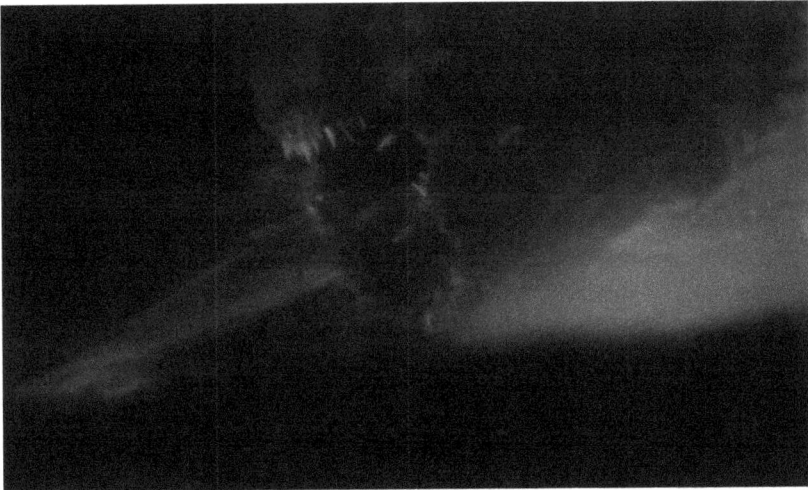

Figure 3. A modern-day Leviathan in *Underwater*, directed by William Eubank (20th Century Fox, 2020).

in *Deep Star Six*) does not really cause us to rethink our place in the cosmic order of things.

But when Leviathan is truly being invoked, the horror is about what the monster says about the universe, and our small and tenuous place in it. If Leviathan had not already existed in thousands of years' worth of world mythology, then Lovecraft would have been the one to invent him. I would offer two recent films that move towards this Lovecraftian cosmic horror through their sublime sea monsters, and, as a result, manage to get in touch with Leviathan: *Underwater* (2020) and *Sea Fever* (2019). While *Underwater* features an impressively Lovecraftian Leviathan at its climax both in scale and it's extra-terrestrial origins, I will focus on *Sea Fever* as the recent film in which Leviathan is most fully present.

Rather than being set in an underwater laboratory, as many of the other films mentioned, *Sea Fever* follows a deep-sea fishing boat on an ill-fated journey. The first sign of trouble is when we realise that Captain Jack has departed from their set course, and instead taken the boat into an "exclusion zone" in hopes of hauling in a bigger catch. Literally, the exclusion zone indicates an area where fishing is not allowed; but additionally, it implies that the crew have wandered into a territory where they were not meant to be. As we will shortly discover, they have entered the domain of Leviathan.

Initially, the crew notice odd holes appearing in the ship's hull; they eventually figure out that something has attached itself to the ship and is eating away at the hull from the outside. Siobhán, the scientist tagging along with the expedition, suggests that they might be some strange species of barnacles, and offers to go diving to get a better look at them. She quickly realises they are tentacles, belonging to the same massive creature – one which she glimpses briefly in the depths before returning to the ship in terror.

For most of the film, the crew deals with the fallout from parasites that this tentacled slime has introduced into the ship's ecosystem. But it is this quick glimpse of the deep-sea Leviathan that lingers. It is an image of the sublime; overwhelming in a way that the design of so many other sea creatures are not. In this image, Siobhán saw a glimpse, as do the viewers, of "the scratching of outside shapes and entities on the known universe's utmost rim," in the words of H. P. Lovecraft (1973: 16). But in keeping with the venerable horror tradition of implying more than is directly shown (and likely also in keeping with

Figure 4. Glimpses of the sublime in *Sea Fever*, directed by Neasa Hardiman (Signature Entertainment, 2019).

the film's limited special effects budget), this brief glimpse of the Leviathan creature is all that is shown, either to the characters or to the audience. For the rest of the film, this monster reveals itself only in hostile microorganisms and unsettling trails of ocean lights.

In the middle of God's speech about Leviathan, God asks Job, "Were not even the gods overwhelmed at the sight of it?" (Job 41:9). As Siobhán experiences with her own glimpse of the deep-sea monster, the very presence of Leviathan is enough to shake an individual to their core. This is not the only instance in which the connection of Leviathan and sight is emphasised in *Sea Fever*. As the tentacles penetrate the ship, they leave microbe-filled slime behind them; it is too late when the crew realise these microbes have entered their water supply, and from there can pass into their bodies. The symptoms progress quickly and are gruesomely fatal. At first, the infected crew member displays symptoms of madness; Siobhán first realises something is wrong when her companion decides it is a perfect time to go swimming, in the middle of the night, in the middle of the ocean. Soon after this, Siobhán observes flecks of green swimming beneath a crew member's cornea, a precursor to his eyes, themselves, exploding. This crew member's sight is "overwhelmed" in the most literal of ways – not having even caught a glimpse of the beast's body, as did Siobhán, the small pieces of the creature he came into contact with were enough to destroy his capacity for sight.

Another key element of *Sea Fever* brings it into the realm of the Leviathanian sublime. In *Sea Fever*, the goal is not to defeat the monster, but merely to escape. In the other mentioned sea monster movies, the creatures are dispatched at the film's climax, often in a deeply silly manner. The crustacean of *Deep Star Six* is undone by gasoline and a flare gun; the beast of *Leviathan* has a demolition charge lobbed down its gullet. Even in *Underwater*, the delightfully Cthuluesque creature is destroyed by a nuclear explosion. But in *Sea Fever*, there is no attempt to fight this creature. It is not even a consideration for the crew. Instead, they know their only hope is to get out of the exclusion zone and back into safe waters. The sense of cosmic horror instilled by Leviathan precludes anything like a happy ending; having the beast explode, and the protagonists sail off into the sunset, instead presents a world in which challenges can be overcome and threats to the kind of "normality" described by Robin Wood (2018: 83) are defeated. The order of the universe has not been disrupted. But in *Sea Fever*, the crew only hopes to escape from the beast's notice. They find themselves in a similar position to the characters in Algernon Blackwood's tale of cosmic horror "The Willows," in which a pair of companions on a canoe trip makes camp on an off-limits island. Realising the terrible situation they are in, one of them holds out a sliver of hope: "Our insignificance perhaps may save us," he says (Blackwood 2002: 51).

In God's speech to Job, God makes a mockery of the idea that a human being would attempt to conquer Leviathan. "Can you fill its skin with harpoons," God asks rhetorically, "or its head with fishing spears?" (Job 40:7). To drive the point home further, God asks Job, "Who can confront it and be safe? – under the whole heaven, who?" (Job 41:11; NRSV). Leviathan is the primordial chaos monster, whose only worthy opponent is one of the gods. Any monster that can be brought low by humans, even with the aid of a nuclear blast, loses its claim to this lineage. And even in the creation myths which depict Leviathan as being defeated by one of the gods, there is always the sense that Leviathan will return. This is perhaps most clear in the biblical tradition, in which the "twisting serpent" (Isaiah 27:1) of the primordial battle of creation returns as the "ancient serpent" of Revelation, named as, "the Devil and Satan" (Rev 20:2; van de Kamp 2017). Only a god may defeat Leviathan, and even then, only at the beginning and the end of time.

In recent fiction, the cosmic sublime of Leviathan is perhaps best encapsulated in John Langan's novel *The Fisherman*. The titular figure at the heart of the novel is an ambiguously supernatural walker-between-worlds, who is either working to keep Leviathan bound so as to stave off the apocalyptic battle or in an effort to gain the power of the monster for himself. The characters are confronted with a cosmos far larger and more horrifying than they can comprehend and are left chillingly aware of their infinitesimally small place in the universe. "There is so much of it that its very presence presses on Jacob," the narrator tells us "as if mere proximity to it might be sufficient to snuff him out, like a candle in a hurricane" (Langan 2016: 146).

Being the hallmark of cosmic horror, the figure of Leviathan should make us reconsider our humanity and shake up our assumed knowledge of the cosmos. God's purpose in describing the awesome power of Leviathan to Job was to push Job to rethink the centrality of his experience; rather than viewing himself and his suffering as the factors around which all of creation is centred, Job's confrontation with Leviathan forces him to reconstruct his view of the universe, a view in which humanity is not central. Similarly, cinematic Leviathans should leave us wondering if we really are at the top of the food chain, or if we only persist as a species because we have, until now, evaded the notice of the cosmic forces which are so much greater than us. When a film successfully evokes Leviathan, these disquieting feelings are not chased away by gasoline and a flare gun. They persist after the credits roll, because we have been brought face to face with the disconcerting reality that the only way to survive an encounter with Leviathan is to escape its notice.

Debaditya Mukhopadhyay

The Matsya Avatar (300 AD–present)

Indian Myth

A tale popular in the Nicobar Islands featuring a mermaid married to a human represents the Indian counterpart of European merfolk legends as listed in *The Penguin Book of Mermaids*. Though this tale's reversal of the traditional plot-line of mermaid–human love story looks unmistakably Indian for its similarities to an episode from *Ramayana*, its representation of the mermaid as a beautiful wife whose story ends with a broken marriage does not differ dramatically from the European mer-wife tales such as Melusine (d'Arras 1382–94).[1] Similarly, the tale of *Suvannamaccha*,[2] featured in versions of *Ramayana*, echoes other tales that feature sinister mermaids who seduce and kill their male victims. While these mer-tales represent the appropriation of mer-legend tropes popular outside India, they do ask the question: do merpeople only appear as beautiful maids in the Indian popular imagination? This chapter will analyse a comparatively less highlighted yet more ancient Hindu myth of the *Matsya* (fish), an Avatar of God Vishnu and its declining afterlife in order to explain how the Indian popular imagination fashioned a merbeing notably different from traditional mer-legends, why this myth's portrayal of human–non-human

---

1   In Ramayana, Sita, the wife of Rama, asks him to go and bring her the golden deer and it is this request for the golden deer that leads to her separation from Rama. Similarly, in "Shoan, a Nicobar Tale" (see *The Penguin Book of Mermaids*), the mer-wife loses her husband when she requests him to bring her a mirror.
2   In a number of Southeast Asian versions of Ramayana, this mermaid-demon appears as an offspring of the demon king Ravana who tries to digress Hanuman from his mission under the instructions of her father (see Sastri 2006).

relations differ from majority of mer-legends, and finally, what has subsequently obscured this myth and the figure of the Matsya Avatar.[3]

Formation and Transformation of the Myth: An Overview

There is no single Theogony for Hindu myths. Rather, the myths appear in the form of multiple Puranas. The Sanskrit word "Purana" literally means "old", but in the context of Hindu mythology, the word refers to a vast body of narratives featuring popular Hindu deities. According to Dimmitt and Buitenen, Purana, as a genre, started coming together since 300 AD and as a whole it documents the changing popularity of various Hindu deities (1978: 3). Alternatively, Vettam Mani suggests Puranas to be pre-Christian narratives that form the very foundation of Indian worldview and traces the word's etymological roots in the combination of the words "Pura" (old) and "Nava" (new) following the scholar Rangacharya (1975: 617). Myth of the Matsya is a part of the Puranas centring round the Hindu god "Vishnu" and is available in three interrelated versions. Though Mytheme-wise the tale is not about a mermaid but a fish and a human being helping each other, in each of the three versions, this fish is revealed to be a form adopted by Vishnu for specific purposes. Instead of outlining these chronologically, this discussion will place these versions following an order that reflects the myth's accumulation of details.

Myth of the Matsya appears in its most rudimentary version in *Shatapatha Brahmana*, which is not a Purana per se and yet is functionally similar to the Puranas to a great extent. In this version Manu, the first man, finds a fish while washing himself (there is no clear mention of the time this incident took place but Manu being the first man, it can be argued to have taken place at the very beginning of things) and is requested by the fish to help it survive and in return he will get saved from a great deluge. After growing into a creature of

---

3   I am indebted to a lecture delivered by Professor Lajwanti Chatani of Maharaja Sayajirao University of Baroda for my analysis of the gradual obscuring of *Matsya* avatar myth.

significant strength, the fish takes leave of Manu leaving instructions for saving himself from the flood. Finally, the fish carries Manu's ship through the great flood by its horn and saves him while the entirety of creation is wiped out.

A more detailed version appears in *Matsya Purana* (see Joshi 2020) which, as the title suggests, is specifically concerned about the glory of the Matsya avatar. In this version the man saving the fish is a pious King named Vaivasvata Manu, who despite being a worthy ruler decides to start tapasya (penances) like a sage. When the god Brahma tells the King to ask for a boon, he is pleased by his response when he humbly asks to become the saviour of creation. Even after earning Brahma's blessings, the King continues to live like a sage and one day while offering prayers to his ancestors, he finds a tiny fish in the cup he was using. Out of kindness he preserves the fish in his water-pot. Subsequently, the fish grows very fast and begs the King to save it when its size exceeds the container used for its preservation, to which the King responds by transferring the fish into a pitcher, then a well, a tank, the river Ganga and finally into the sea. When the fish's growth seems to exceed even the sea's vastness, the King becomes afraid and requests the fish to reveal its true nature. In response the fish confirms that it is none other than Vishnu and warns the King about the imminent flood. He then gives the King the duty of building a boat in which he will have to save all living creatures from the flood by sailing through the deluge tying the boat to the great fish's horn.

Apart from presenting the King's kind nature in greater detail, by way of depicting how he goes on helping the fish's survival during its growth without any expectations, *Matsya Purana* also presents the Matsya as an epitome of knowledge. While the first two chapters of the Purana outline the basic storyline by showing Manu's meeting with the Matsya, the avatar revealing itself, and the beginning of the great flood, the remaining chapters (more than two hundred and fifty) of the Purana continue narrating how Matsya shared its profound knowledge about the unfolding of Hindu pantheon with its dedicated listener King Manu. In so doing, the Purana presents Matsya not just as a saviour but also as a protector and progenitor of knowledge and culture. It is worth mentioning that a similar idea is found in *Bhagavata Purana's* account of the Matsya Avatar. While *Matsya Purana* does not discuss exactly when and how Vishnu became a fish, in *Bhagavata Purana* it is mentioned that Vishnu had to transform himself into a fish when the Vedas were stolen by a demon named Hayagriva (Menon 2012: 522). Overall, the Matsya Avatar appears as a true guardian of creation as well as knowledge and culture and it is this last aspect of the Matsya Avatar that draws it significantly closer to the figure of a merperson.

Is Matsya a Mer-Legend?

In a study focused on *Matsya Purana*, the Indian scholar V. R. Ramachandra Dikshitar drew parallels between Matsya Avatar and "the legend of Oanness" highlighting how both Matsya and Oanness enlighten their listeners with ancient knowledge (1935: 4). Looking closely at the figure of Oanness, this section will explain how significantly Oanness and, by extension, Matsya resemble European merpeople. Sir Arthur Waugh describes Oanness as "the first merman in recorded history" (1960: 73), referring to Oanness' description by Polyhistor. As per Polyhistor's account, Oanness, who has features of both the human and piscine anatomy, enlightened Babylonian people with knowledge of both science and arts in a human voice (qtd. In Waugh 73). From the image given below (Figure 4), the similarities between Oanness and the merpeople as known to us at present become more evident.

Figure 5. Image of Oanness with noticeable Mermaid features. Image in public domain.

The borrowed nature of the Matsya myth has been highlighted by Wendy Doniger O'Flaherty who argued that the myth of Matsya was not "originally associated" with the Hindu god Vishnu (1975:179), instead the myth owes its origin to "Semitic flood legends" (180). Interestingly, the legend of Oanness is also of similar origin. As explained in the *Dictionary of Deities and Demons* in the Bible, the name "Oanness" was actually derived from the name "Uanna" who was "the first" of the seven (1999: 73) legendary beings of Mesopotamian religion known as *Apkallu*. Though these seven *Apkallu* differ from Matsya for their prediluvian nature, taking into consideration the multiple similarities between these Babylonian legends and the narrative of Matsya Purana, it seems plausible to view the myth of Matsya as a conglomeration of Babylonian legends that brought about a mergod, if not a mermaid, for the Hindus.

While the account of the Matsya myth in *Matsya Purana* connects the figure of Matsya with merpeople by way of laying bare its connections with ancient legends, the version found in *Bhagavata Purana* links the two by pointing out Vishnu's shape-shifting abilities. As mentioned above, Vishnu had to transform into a fish on an urgent basis due to the crisis brought about by the stealing of the Vedas. Since most merpeople are considered to be effortless shape-shifters (Bacchilega and Brown 2019: xiv) and three of the ten Avatars of Vishnu (Matsya, Kurma/Tortoise, Baraha/Boar) show the god changing into various animals, it seems tempting to view Vishnu's transformation into Matsya Avatar as an act of shape-shifting similar to certain legendary merpeople. Further, iconographical evidence suggests a strong resemblance between the Matsya avatar and merpeople. As noted by Nanditha Krishna: "Matsya is generally represented as a four-armed figure with the upper torso of a man and the lower torso of a fish. *Only occasionally is he represented as a full fish*" (emphasis added Krishna et al. 2018: 64). The images represent the worshipped idol of Matsya Avatar from Sri Matsya Narayana temple, Karnataka (Figure 5), and a sculpture featuring the Matsya Avatar found in Chennakeshava temple (Figure 7).

Figure 6. Use of human face in the idol from Sri Matsya Narayana Temple. Image in public domain.

While the differences between the two images, namely the use of a piscine face in Figure 7, indicate the differing ways ancient Indians imagined this Avatar, in both the images the co-presence of human and piscine attributes, which is a key feature of the merbeings, is prominently visible.

Figure 7. Use of the piscine face in the idol from Chennakeshava Temple. Image in public domain.

Uniqueness of the Myth

Being a manifestation of humanity's collective response to aquatic life forms, mer-legends across the world traditionally centre round a number of tropes and stereotypes through which these "reflect an anthropocentric view of the world" (Bacchilega and Brown 2019: xix). In comparison with these traditional mer-legends, the myth of Matsya presents a tale highlighting the importance of human–non-human co-existence. Two among the three recurrent plot-patterns of mer-legends across the world, as identified by Bacchilega and Brown (xvii), depict the merfolks' relationship with humans to be unnatural and avoidable to both sides. The plot-pattern involving the mermaid's transformation into a woman is fraught with anxieties about the incompatibility between the two species. Similarly, the tales featuring "humans being held captive underwater" (Bacchilega and Brown 2019: xix) maintain similar views about this relationship by refraining from showing the exact experience of the captive human among the merbeings. Even mer-legends representing merfolks and humanity "interacting temporarily in contact zones" (Bacchilega and Brown 2019: xvii) do not actually differ from the former two plot-patterns due to their portrayal of hierarchical relationship between the two groups.

Matsya Avatar's tale differs from the above patterns by presenting Manu and the Avatar in a symbiotic relationship. In all the versions, Matsya and Manu help each other to survive, thereby offering a narrative of co-existence. Though the role played by the Matsya during the flood or its divine nature apparently puts the Matsya above Manu, the depiction of Manu's unconditional kindness to the Matsya in its small and helpless state places Manu, and by extension humanity, in a position of equal importance. Besides, the interactions between Manu and Matsya, the imparting of knowledge by the Avatar during the deluge, shows the human–merbeing encounter to be an enriching experience for the human, instead of being an accidental or forced meeting. As a whole, the Matsya-myth, when viewed as a mer-legend, represents a remarkable subversion of the anthropocentrism prevalent in mer-legend traditions. The figure of the mermaid, which is arguably "the central figure of the merfolk", remains absent from the myth of Matsya, but the myth's distancing from this oft-stereotyped figure actually widened the narrative's scope by ridding itself of the necessity of plot-patterns featuring seductive or lovelorn mermaids. Conversely, the absence of the mermaid figure and the countering of anthropocentric worldview have turned out to be detrimental for the afterlife of the Matsya narrative.

The Myth Now: An Obscured Presence

At present the myth of Matsya has largely turned into an obscure legend as it is at odds with prevalent cultural discourses. On the one hand, the gender-neutral figure of the Matsya, as found in Indian mythical narratives and iconography, has made it difficult to be appropriated into the cultural network influenced by European tradition of mer-legends which prefers to sexualise merfolk through the figure of the mermaid or siren – even mermen need to be hypersexualised as in *Aquaman* (Wan, 2018), for example. On the other, even the cultural imagination of the Hindus that initially treated Matsya to be a divine entity has largely ignored updating this myth due to the prevalence of anthropomorphism in Hindu iconography. While other Hindu deities with non-human attributes like Ganesha or Hanuman have achieved considerable popularity by regular representation in television series or animations and have achieved a dominance of the human form within anthropomorphism, which in the case of Matsya is comparatively difficult to achieve.

Lastly, it is significant that *Moby Dick*, arguably the most popular appropriation of the figure of the Matsya Avatar (see Sullivan and Hall 2001), shows humanity desperate to slaughter the mighty aquatic entity instead of co-existing with it. The ancient mergod is indeed too subversive for this anthropocentric era of vilifying or sexualising merfolk.

Manal Shalaby

Scales (Shahad Ameen, 2019)

There are two coastal cities in the Middle East nicknamed *arous al-bahr* (Arabic for "maiden of the sea"): Alexandria in Egypt and Haifa in Palestine/Israel, both overlooking the Mediterranean Sea; hundreds of seafood restaurants across the region also use *arous al-bahr* as their business name; and mermaid-themed plushies, dolls and school supplies are a familiar sight in Middle Eastern toy stores and stationeries. The legend of the creature that has the upper half of a beautiful woman and the lower half of a giant fish is known to almost everyone in the Middle East despite the fact that this knowledge is largely attributed to the universal popularity of Hans Christian Andersen's *The Little Mermaid* (1837). This does not mean that the region's mythological and folkloric heritage is devoid of merfolk stories. The Babylonian Oannes, the Sumerian Ninkharsag, the Ancient Egyptian Isis and the Assyrian Atargatis are all deities closely linked to water, fertility and water-land hybridity. Besides, Middle Eastern historians and scholars such as Al-Mas'udi (896–956 AD) and Al-Qazwini (1203–1283 AD) relate in their cosmographic writings fantastical accounts of sailors encountering sensible fishlike creatures, quite similar to the ones we may find in *One Thousand and One Nights*[1] (also known as *The Arabian Nights*).

Notwithstanding the rich textual and oral body of merfolk stories in the Middle East, later cultural productions seem to be heavily influenced by Western retellings of mer-myths. In the second half of the twentieth century, audio-visual productions of the mermaid legend were favourable choices to

---

1   These tales include "Julnar the Sea-Born," "The Adventures of Buluqiya" and "Abdullah the Fisherman and Abdullah the Merman."

production companies and audience alike due to the growing influence of Hans Christian Andersen's story which by then has made its way to children's literature published in Arabic. For instance, the 1985 Egyptian series *Alf Leilah w Leilah* (*One Thousand and One Nights*) features an episode titled "Arous al-Bohoor" ("The Maiden of the Seas") which, ironically, has nothing to do with the merfolk stories mentioned in *One Thousand and One Nights* and bears more affinity to Andersen's beloved tale. Seven years later, "Arous al-Bahr," an episode of the Syrian television series *Kan Yama Kan* (*Once Upon a Time*) (1992) tells the story of a poor fisherman who catches a mermaid in his net. The mermaid promises the fisherman to grant one of his wishes in return for releasing her into the water. The fisherman releases her, and she lives up to her promise; however, the fisherman's greedy nature urges him to keep asking the mermaid to grant him wishes until he is forced back into his initial impoverished lifestyle. The story's plot-line is unmistakably evocative of the Brothers Grimm's well-known fable "The Fisherman and His Wife" (1812) in which the poor fisherman and his wife are also eventually punished for their greed by an enchanted fish whom the fisherman releases from his net at the beginning of the story. Although both productions are direct adaptations of Western texts, they still reflect some of the religious and traditional qualities pertinent to Middle Eastern artworks at the time, such as adherence to family values of decency and modesty, foregrounding moral lessons and religious teachings, and observing the oral tradition of rhyming speech and/or using standard Arabic.

With the advent of the twenty-first century, those outcrossed productions did not seem to endure as artists have started to redirect their gaze towards earlier sources of Middle Eastern mer-myths for inspiration – one such artist is Saudi filmmaker Shahad Ameen. In 2013, Ameen wrote and directed a 16-minute short entitled *Houreya wa ain* (*Eye & Mermaid*) which she, a few years later, developed into the feature-length film *Sayyedat al-Bahr* (*Scales*) (2019). Both films tell the story of an unspecified island whose people live in tribal conditions and feed on the flesh of the mermaids they catch; while the two films share the same backstory and central characters, they diverge into different directions that can be seen linearly complementary. *Eye & Mermaid* introduces us to the spirited 7-year-old Hanan (played by Basima Hajjar) who sneaks to her father's fishing boat behind his back. After discovering her presence, he reproaches her lovingly saying that the sea is not for girls, and then gives her a black pearl for her necklace. On land, Hanan surreptitiously follows her father and his fellow fishermen as they carry a screeching mermaid to be butchered, and it is then that Hanan finds out that her cherished pearls are brutally cut out of the mermaid's skin. The frightened child feels pity towards the now dismembered mermaid and decides to free her by dragging her body back to the sea. At the end of the film, an adult Hanan sails her deceased father's fishing boat all by herself and then jumps into the water where she meets the mermaid and hands her the black pearl necklace.

The story's dynamics change dramatically in *Scales* as the islanders follow an age-old ritual in which they sacrifice their female firstborns to the sea to insure bountiful fishing harvests (of mermaids). The film opens with a father in the act of sacrificing his infant daughter, but he hesitates and pulls the crying baby out of the water as it was about to be snatched by a scaly hand. A few years later, the baby grows up to be the 14-year-old Hayat (also played by Basima Hajjar), still haunted by her near-death experience and the indignation of the entire tribe. After her mother gives birth to a boy, Hayat's father is forced to sacrifice his teenage daughter once again. Hayat bravely ventures into the sea, disappears for a few hours, and then, to everyone's amazement, comes back dragging the body of a mermaid. As a result, the head of the fishermen welcomes her amongst his men and she joins them on their fishing trips. On one of those trips Hayat discovers the horrific truth that the mermaids they catch, albeit having fish tails and covered in scales, are no other than the girls sacrificed earlier by the tribe members. Hayat jumps into the water, not to be seen for a few days during which the sea supernaturally disappears, and the islanders face the imminent calamity of drought and famine, only to show up with another mermaid carcass. This time, however, Hayat gives the mermaid a proper burial and forbids future fishing and eating of mermaids. She cries, and when her tears touch the barren seabed, the water comes rushing back.

Although many narrative and technical elements set the two films apart, it is hard to discuss them separately. Ameen has woven two stories together in order to constitute one dialectic narrative constantly borrowing from and building on itself. Hanan/Hayat has to confront a male-dominated society that not only marginalises its female members but literally devours them. Hanan first comes face to face with the brutality of this society when she watches the fishermen holding the struggling mermaid down, gouging the pearls of her skin and trying to cut her into pieces in what looks like a gang rape scene (Hayward 2016: 2). Hayat, on the other hand, moves from the spectator's seat to the victim's seat as she realises that the mermaids the fishermen violate and eat are actually the young women of the tribe, herself included. The realisation places Hayat inside the horrendous cycle of subjugation which Hanan only witnesses from the outside: the cycle begins with alienating the female members of the tribe by limiting them to the birth-giving function, then using them as sacrificial offerings for the benefit of the tribe, and finally labelling them as prey to justify the alienation and violence committed against them all along.

The concept of sacrificial human offerings in Hayat's world bears strong resemblance to ancient Middle Eastern mer-myths in which female deities' relation to the water is also marked by an element of sacrifice. The Assyrian goddess of fertility Atargatis, who was often depicted as a mermaid, was believed to have relinquished herself to the water after causing the disappearance of her mortal love interest, which resulted in her piscine metamorphosis (Coulter and Turner 2012: 78–9). Like Atargatis, the young women in *Scales* are forced into the water, even though not driven by their own sense of guilt, and they naturally blend into the new element and metamorphose into half-woman half-fish creatures – this holy integration that the Assyrian goddess has earlier achieved is nevertheless regarded by the men of the tribe as an abominable transgression against nature. Isis, goddess of fertility and rebirth and Atargatis' Ancient Egyptian equivalent, is another source of inspiration for Ameen's mer-myth. Egyptian myths tell us that the bountiful flood of the Nile is caused by the tears Isis sheds over her dead husband and brother Osiris; consequently, Ancient Egyptians would honour her every year by throwing an effigy of a maiden into the Nile to herald the new flood season and to allow Isis to reunite with her beloved husband (Gadallah 2016: 118–19). Ameen turns the effigies into real maidens in her story, but she bestows upon Hayat Isis' divine power of commanding the water. The tears Isis cries over her dead husband are the same as Hayat cries over the dead young girls; they both entail a great sacrifice but, in return, bring about the life-affirming tides of water.

Hanan's development into Hayat is a coming-of-age story – one of resistance and resolve. Hanan does defy the tribal rules by returning the mermaid to the sea, but her relationship with the self-reflective image of the mermaid

remains binary like the relationship between the two nearly identical protagonists of *One Thousand and One Nights* stories "Abdullah the Fisherman and
Abdullah the Merman." In this allegorical tale, a fisherman named Abdullah
encounters a merman, also named Abdullah, who lives a life similar to his but
underwater. Throughout the tale, and despite their failure to reconcile their
differences, the two Abdullahs stand as an image of the Self and its reflection;
each can identify with and see himself reflected in the other without ever fully
becoming him. Similarly, Hanan relates to the mermaid's suffering, but they
eventually remain two separate entities, each belonging to her own element/
world (Figure 8). Finding the idea of self-reflective images no longer adequate,
Hayat, the older Hanan, takes the more difficult road towards understanding
and self-actualisation. Hayat first revolts against the role of the passive victim
and directs her anger towards the other victims, that is, women of the tribe and
mermaids. Therefore, she further alienates herself from her peers and kills one
of the mermaids to prove her worth and agency to the patriarchal society she
is part of, which leads her to join the group of men and to be treated as one.
It is true that Hayat refuses to be victimised, but her rather adverse reaction
backfires and she suddenly finds herself assuming the role of the abuser who
instigated her rebellion in the first place. Although she has earned her place
among them, Hayat finds herself the object of ridicule at hands of the fishermen
who attribute their failures and bad luck to Hayat's feminine presence. Finally,
Hayat realises her mistake in thinking of the mermaids as a mere exotic reflec-
tion/projection; her guilt and shame move her to jump into the water – just
like Atargatis did – and when she emerges, she comes to terms with her fluid
and comprehensive identity (Figure 9). Only then does she assimilate all parts
of herself (human-mermaid, land-water, natural-supernatural, female-male)
and asserts her individuality, becoming the divine force of life itself. It comes
as no surprise that *Hayat*, in Arabic, means "life." Hanan/Hayat's process of
integration is also semantically indicated in the films' titles: whereas *Houreya
wa Ein* (*Eye & Mermaid*) clearly refers to the existence of two separate per-
sonas, *Sayidat al-Bahr* (*Mistress of the Sea*, original Arabic title of *Scales*) observes the protagonist's transfiguration into one unified entity.

The unconventional state of wholesome in-betweenness that Hayat enjoys
is not only physically manifested in the black and white harmoniously dominating the film's duo-chromatic image or the scales covering Hayat's feet

Figure 8. Grown-up Hanan comes face to face with her exotic piscine counterpart in *Eye & Mermaid*, directed by Shahad Ameen (Doha Film Institute, 2013).

Figure 9. Two becoming one in *Scales*, directed by Shahad Ameen (Variance Films, 2019): Hayat's scaly feet flow into the mermaid's piscine identity, while the mermaid's braided hair – identical to the braids Hayat wore earlier – flows into Hayat's human identity.

or in her ability to traverse land and sea effortlessly; this state envelopes the entirety of Ameen's mer-myth and its ability to negotiate the boundaries between fantasy and reality. Usually, children and adults alike tend to resort to fantasy to fill in the gaps of their knowledge of the world and create concepts that occupy medial positions such as werewolves (between "man" and "wolf") and mermaids (between "woman" and "fish"). Mermaids do exist in Ameen's story as a fact; however, it is Hayat herself who employs her gap-filling faculties to create a space for herself between the victimised mermaids/women and the abusive fishermen/men. Through her process of self-actualisation, Hayat mends the gap between the two opposing sides and transforms into the very myth she needs in order to become whole.

Shahad Ameen's culturally authentic tale condemns the discriminatory patriarchal rules and traditions women have to reel under in some Middle Eastern countries by incisively drawing inspiration from the rich folkloric heritage and the female-revering mythology of the same countries. The tale, nonetheless, does not stop at that. In an interview with *Variety*, Ameen says: "Males write history, and they write the cycle of this world … I wanted to tell a timeless story because it could happen anytime and, really, any place" (Vivarelli 2019). Hanan/Hayat is a young girl who struggles out of the victim's role in a world rife with vehement political, religious and social dissensions, but she is also Isis and Atargatis – the timeless reminders of human beings' capability of congruous assimilation and of women's life-preserving powers. In that sense, the nearly generic narrative worlds of *Eye & Mermaid* and *Scales* seem to transcend the specificity of a certain time, place or religion to address the divinely mutable nature of femininity and the question of human identity at large.

Amylouise Ahava

The Great Horned Serpent
(c. 1450–present)

Iroquois Myth

Locals and tourists of any lake shore will at some point, or another tell tales of a monster which hides just below the surface. Some of the "greats" are Champy of Lake Champlain in the Adirondacks, and obviously, Nessie of Loch Ness in Scotland. Both legends involve a plesiosaur-looking creature who now serves as a local mascot and adds to the tourism of the area. The monsters are well known, yet the creatures do not have a backstory and their storylines do not develop much beyond a giant dinosaur-like creature in a lake. However, stories from the Indigenous people of North America put a lot more care and importance into developing their lake monster myths. And when explaining North American folktales and myths, the Iroquois[1] "rank among the most imaginatively rich and narratively coherent traditions" (Wonderley 2009: xiii). Straying from European and European-influenced American lake monster myths, the creatures of the First Nation legends are more similar to Asian myths, plus the creatures have ulterior motives and possess far more intelligence and cultural importance than an average lizard. For many Indigenous storytellers the plots and themes associated with the water monsters give a glimpse into the history, culture and psychology of the local Natives.

In this chapter I will focus on the Great Horned Serpent[2] of Iroquois folktales which comes from stories dating back hundreds of years. The stories mostly centre on the Great Lakes of North America, where folklorist George

---

1   The Iroquois Confederacy is a collection of six tribes (originally five) which take up residency around the Great Lakes of North America.
2   The Great Horned Serpent goes by the name of Oniare (also spelled Onyare, Onyarhe, Ohnyare), Oniont, Oneyont, Agont, Donogaes, Doonoaes, and Jodi'gwadon.

E. Lankford (Wonderley 2009: 52) believes a lake deity once existed as a wide-spread belief. He refers to the creature as the "Great Serpent" from the land of the watery beneath which is almost always described as a Great Horned Serpent (52). The origin of the stories is difficult to trace because interactions between tribes caused adaptations so the tales would fit different narratives. The description of the serpent varies with some stories mentioning poisonous breath, while others bring more attention to the monsters' great horns. The Great Horned Serpent's horns sometimes earn a description similar to the massive antler rack of a white-tail deer, and other times the horns mimic an exaggerated pair of bison-horns. Collections of these tales come from a wide variety of sources; therefore, the retellings of the Iroquois myths for the sake of this paper might deviate a bit from the original Iroquois rendition of the monster or the versions more familiar to some Native readers. Furthermore, many of the stories largely remain in the oral traditions of tribal Storytellers or older relatives. I, myself, grew up hearing stories of Oniare from my grand-father who used the cautionary tale of a giant lake snake to keep me from going to the end of the pier or from making too much noise while out on the water. Hearing (in great detail) about the horrific creature no doubt lurking just a few feet below our boat created a strong fear of water creatures I still have to this day.

Numerous Native myths changed over time due to interactions with different tribes, but the largest alterations to the stories came from the white European settlers. Myths of lake monsters, especially those from the United States and Canada, are often published only after being discovered by colonisers and therefore offer a perspective much different from that of the local tribes. Some of the traditional lake serpents over time grew to resemble four-limbed lizards of European folktales rather than the snake-like creatures common in First Nations tales. These creatures mimic dinosaurs, crocodiles and even the mythical dragons. Perhaps the difference in appearance between European and First Nations Lake monsters comes from the ubiquitous images of the plesiosaur as the discovery of the dinosaur occurred in the 1820s and the influence of the dinosaur directly impacted the description of other large water-dwelling lizards.

Another reason the colonisers strayed from the original image is because the European snake represents the "bestial unconscious" and the ultimate

Christian villain: the Devil. In fact, the artistic or literary portrayals of snakes in European tradition depict serpent-like creatures as adversaries whom the protagonist must "master, carve, or skewer" (Bastine and Winfield 2011: 225). In more recent times the Indigenous visions of the lake monsters became even further removed, and due to TV, movies and even comic books, the modern stories have left out Native voices altogether. Nowadays, many of the modern North American versions use the creature to sell soda, draw in tourists, or serve as a mascot for the local little league team. Champy's original incarnation is described as a serpent, but after 1978 it is transformed into a dinosaur-like monster, which sceptics and investigators Benjamin Radford and Joe Nickell attribute to the popularisation of Scotland's Nessie and the familiar shape of the plesiosaur (Radford and Nickell 2006: 40). In a recent episode of the Shudder TV series *Creepshow* "By the Silver Water of Lake Champlain" a two-generational search for the creature finds a monster strongly resembling Nessie and not the original Native version (see Figure 10).

In traditional Iroquois tales, the appearance and purpose of lake monsters hold a stronger resemblance to Asian stories because in both cultures the snake/dragon image maintains a strong connection to the earth and the "power

Figure 10. Champy's dead baby shows a long neck and plesiosaur-like flippers; from "By the Silver Water of Lake Champlain," *Creepshow*, directed by Tom Savini (Taurus Entertainment Company, 2019).

of fate" (Bastine and Winfield 2011: 225). Instead of conquering the creature, the hero of the story must find a way to work with the monster. Despite some cultural influence from Christian Europe, the Iroquois serpent folktales "have a bit of both continents in them" (Bastine and Winfield 2011: 225). Not only does the physical appearance more closely align with Asian lake monsters than with European, but the snakes' abilities and their background stories also hold a strong resemblance to the Asian legends. In myths from Asian and Native culture there appears a common theme of the connection between thunder and lake monsters. The thunder above and the creatures below the water are an essential myth and these themes of water monster versus sky appear "deeply rooted in regional cosmology of a former time" (Wonderlay 2009: 52). For example, on Mount Chiok in South Korea there is a pond near a Buddhist temple which used to serve as the home for nine serpent-like creatures that used thunder to alter the landscape and scare away the monks (Darkside of Seoul Podcast n.d.).

And even though the Great Horned Serpent of Iroquois tradition did not harness this power, thunder still plays a large role in the stories and the serpent's impact on the Earth. Some tales tell of the creature's attempts to help outcast humans or travellers, while others explain the never-ending battle between the Great Horned Snake and Hinon,[3] the Thunderer (Bastine and Winfield 2011: 227). Sometimes the serpent allows offerings in exchange for the lives of its human prey and other times the thunder god Hinon physically fights Oniare. In fact, many geological features around the shores of Lake Ontario are reputed to be the work of the "climatic clash" between the Great Horned Serpent and Hinon (Bastain and Winfield 2011: 227).

When looking at the characteristics of the Great Lakes, one can easily see why traditions of lake monsters occur so readily. The cold and expansive lakes with no shore in sight almost invite the viewer to conjure up stories of large creatures hidden in their unfathomable depths, or they inspire a spiritual need for a non-human entity to ensure that their passage is a safe one. Celeste Ganassin, curator at the Kelowna Museum, posits that any First Nation group will create an "entity in a lake they had to respect or fear" and these stories

---

3   Also called Hino, Hine, Heno, Henon, Hinu, Hinun, Henoh, Hinen, Heynuh, Hinnon, Hihnon.

Figure 11. The Great Horned Serpent by Jesse J. Cornplanter in *Seneca Myths and Folk Tales* (1923).

existed to explain "storms, sudden wind, and so on" (qtd in Radford and Nickell 2006: 126). In the myths, the serpent lives in the Great Lakes (usually the body of water changes according to the origin of the storyteller) where the creature overturns canoes and eats unsuspecting swimmers or anglers.

In some stories people triumph over the Horned Serpent, but more frequently the creature preys on or offers assistance to an outcast (usually a woman). The woman is an outsider from the tribe and must befriend the creature in order to survive. In the story "The Horned Snake and the Young Woman" the female character serves as a love interest to the serpent and even though the monster does not intentionally want to harm its betrothed, it does wish to remove all traces of her humanity and make her a monster (Hewitt 2010: 269). The Serpent is often charming and mysterious, and it lures the gullible into danger or foolish bargains. In a story told by Seneca Faithkeeper Jesse Cornplanter, "The Orphan Girl: A Legend of the Horned Serpent," an orphan girl, described as "nice-looking" and "industrious," receives poor treatment from the rest of her tribe (Cornplanter 1986: 58). In an attempt to rid themselves of the girl, the tribe tricks her into accompanying the gatherers on the yearly berry-picking trip, only to abandon her on an uninhabitable island. The girl cries and accepts her fate until a voice instructs her to find twelve willow whips and meet it at the shore the next morning. It promises rescue, but she must not be afraid of its appearance. The next day the girl meets her rescuer to discover a giant serpent waiting. It lets her sit on its head between its massive horns and explains she must use the willow whips when he sinks or moves too slowly. He says that once she returns to her tribe they will respect her, but only if she keeps its role in the rescue a secret. The girl and the serpent head across the vast lake and every time the creature tires the orphan whips it for encouragement. The closer they get to the shore, the darker the clouds get and eventually the girl jumps from the serpent's head and swims to the shore because the Thunderer attacks the serpent (see Figure 11). The creature disappears beneath the water, and unfortunately, the girl does not know if it lives or dies. The tribe takes the girl back and holds her with great esteem, just as the serpent promised.

However, the story of the Orphan Girl goes beyond just entertaining because the help the serpent gave the orphan eventually became included in ceremonies and rituals. Cornplanter explains that the purpose of the story is to show how "the things people looked [at] with fear [...] were also their benefactors" (Cornplanter 1986: 58). Even in present day, during the performance of the Great Dark Dance[4] the name of the Great Horned Serpent (Djo-nih-gwo-donh in this case) plays a large role and songs are sung about the creature as a way to "renew the bonds of friendship" with the animal (Cornplanter 1986: 58). The story of the orphan girl does not pass as a bedtime story for children but holds morals appropriate for adults and the tribe as a whole. Cornplanter claims his father spoke of the Great Horned Serpent and similar legends as "reality." "It actually did happen," he says, "but in such a manner as to sound doubtful to us now days" (Cornplanter 1986: 65). Cornplanter agrees with his father because his people lived "so close to nature that it was a sort of common affair to have someone get some help from some animal" (Cornplanter 1986: 65).

---

4   A feast with songs whose intention is the calling of mythical beings.

In oral tradition, the narratives express the belief structure and the overall perception of the tribe. The tales of lake monsters (while on the surface tales of fictional creatures) display "vivid imagery" that conveys an idea of people navigating the richly symbolic landscape of the human condition together. When the orphan girl needs the serpent to take her across the large lake, the body of water represents regular obstacles and struggles in a person's life or even the "journey of the dead to the afterlife" (Wonderlay 2009: 51). Despite the always futile demand for scientific evidence of lake monsters, the search for and discussion of them continues. In relation to the ongoing sightings of Champy, Radford and Nickell (2006: 66) determine that the lasting stories reflect more on the "hunters than the hunted" (66). The reason the Great Horned Serpent or any other First Nation Lake monster continues to exist comes from the "psychological needs" of the human mind and not some environmental mystery. Stories serve as a bond between the past and the present and fill in gaps due to undocumented histories. Exploring folktales allows for the Iroquois to share what makes them culturally unique and everything in the stories, from the creature to the plot lines, explains a part of their culture (Wonderlay 2009: xiv).

Marko Teodorski

Rusalki (Witold Pruszkowski, 1877)

In the dense thicket, above a dead man's body and close to water, the three of them abide. Their traditional Ukrainian costumes are white, accentuated by cinnabar aprons, belts and wreaths. The crescent, inconspicuous moon is hidden high in the canopy. The scene is, nevertheless, illuminated. Their faces are radiant as if infused with moonlight, as if it were inherent to their spectral bodies, immersed deeply in the surrounding reed. The body of water, a lake perhaps, is near, visible low at the bottom, incontrovertibly connected to their presence (see Figure 12).

A faceless man approaches from the distance and a murdered one is lying face down in the shoal; he is half consumed and hidden by the thicket. We cannot see any of their faces, nor recognise them, but we see her – the vanguard – sitting on the branches, body ethereally light, one hand firmly planted in tall grass, face filled with reverie. She could be dreaming of crystal palaces at the river bottom, or contemplating another devilry, since her days on Earth are limited.

She is a rusalka; the men are irrelevant.

All is still.

I led with Witold Pruszkowski's painting *Rusalki* for three, mutually connected, reasons: it is a deep and penetrating gaze into these folkloric beings; it arrests the historic moment of their dramatisation; and it visually articulates a peculiar semantic network of traits I intend to discuss here – paleness, moonlight, sleep/reverie and stillness. Painted in 1877, the scene rests on at least a century-old tradition that saw rusalki both as the embodiments of national sentiment (Emerson 2019: 170; Naroditskaya 2006) as well as a part of a pan-European vogue (see Verba 2017, 2021). France, England and Germany of the

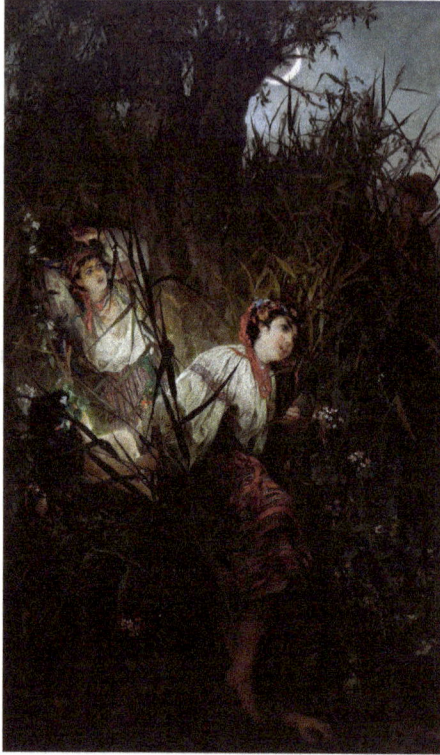

Figure 12. *Rusalki* Witold Pruszkowski (1877). Image in the public domain.

time had been engulfed in images of water maidens who lured men to doom or were tragically spurned by their lovers. Heinrich Heine's Lorelai and other human and half-human sirens (from Thackeray's Becky Sharp to John Collett's Bessie Vane); Hans Christian Andersen's the Little Mermaid and sorrowful fishtailed heroines; the transcendental immortals of H. G. Wells, E. M. Forster and Guissepe Tomasi di Lampedusa; Friedrich de la Motte Fouqué's Undine and mischievous nymphs; Letitia Landon's Melusina, John Keats' Lamia – the list goes on and on. For one thing Victorian men loved above everything else was a female body exuding raw elements, while simultaneously being subdued by their civilising masculinity.

Although they comfortably fitted into the European water maiden narrative, rusalki had the audience of their own. Being mythological Slavic creatures, their image could not compete with, for instance, two-millennial tradition of sirens that sparked with Homer, transformed through the Christian Middle Ages, and was in the eighteenth century finally appropriated by colonial powers as everything classical. For that reason, rusalki predominantly stayed confounded within the cultural framework of Slavic countries, most notably Russia, while remaining part of the wider narrative of malevolence, eroticism and metamorphosis thrown at women by romantic artists (Naroditskaya 2006: 220). From their narrow niche they dialogued with other water maidens assuming some of their traits, yet stubbornly remembering their Eastern European origin. Nikolai Gogol, Ivan Turgenev, Mikhail Lermontov, Ivan Kramskoi and Konstantin Makovsky all wrote about them or painted them, drawing from the deep pool of folk believes; even Pushkin's famous drama *Rusalka* (probably the most Europeanised version of this being) retain perceptibly Slavic traits. Therefore, in order to understand them we need to look at Slavic folklore.

Across the Eastern European Slavic space, from Russia in the north to the Balkans in the south, rusalki are known as female beings associated with lakes and rivers, as well as woods and fields (Vinogradova 2002: 416). Unlike sirens and mermaids (with whom they are often erroneously equated) they are not monsters: their bodies are human, for once they were human beings. According to folklore (but heavily depending on the country, even region, in question), rusalki could have once been unbaptised children, girls who had died during the Trinity week or unmarried (especially the already betrothed ones), or maidens and young women who had violently drowned (Hilton 1995: 143–4; Ivanits 1989: 75; Zečević 1974: 110; Zelenin [1916] 1995: 142). Their pre-narratively victimised state is an important digression from siren and mermaid images – though they do return to drown men or tickle them to death, rusalki have already had drowned themselves. This makes their relationship with victims (men) semantically more complex, because where sirens libidinally lead them elsewhere, rusalki reduce them to themselves. The core of their narrative composition is, apparently, not about luring or destroying the Other, but reducing it to the Same. Therefore, the true plot of rusalki in literature and painting is not seduction, but *revenge* in which the spent libidinal energy is thrown back at the victim. As Caryl Emerson (2019: 174) says about Pushkin's *Rusalka*, "suicide is the pre-story".

Rusalki's appearance is often difficult to pinpoint: depending on the region, they are described as beautiful maidens with long dishevelled hair, naked or in white robes, or as crones, ugly and old women with exaggerated breasts (Ivanits 1989: 76). They can walk among the living during the Trinity week, a specific time of year called *rusalnaya nedelya*, when they pray in woods, near lakes, in fields, or at crossroads, combing their hair, swinging at tree branches and tickling men to death (Vinogradova 2002: 416). During this period, usually May or June, people pay attention not to encounter them and restrain from activities such as swimming in the bodies of water.

But rusalki in literature and art often greatly digress from the ones imagined by folklore. They are romanticised and dramatised, sometimes turned protagonists of their own stories or heavily assuming traits of Motte Fouqué's Undine or Andersen's Little Mermaid. This could be clearly seen in the romanticism of Alexander Pushkin's poetry.[1] In 1819, Pushkin wrote the poem "Rusalka". It tells of a monk living in isolation, praying to God and counting his last days on Earth. One night, as he sits at a lake's shore, he sees a beautiful woman coming out of the water. She combs her "long tresses" (Pushkin [1819] 1888: 84) and frolics; sends him kisses, laughs and cries, all the while beckoning him to follow her into the lake. For two nights the monk resists, but the morning after the third only his white beard is seen under the surface. Pushkin's rusalka is, thus, incredibly close to a siren: she is an aquatic being that lures men with her voice and carries a particularly Christian moral of the story. Just as the medieval sirens were seen as vicious creatures preying on the human soul, the stakes in Pushkin's poem are not only the monk's life but also his faith, ergo his soul. Rusalka is here the embodiment of corporeal sins, a carnal temptress whose body is as sexually deviant as the monk's fate was once pure.

However, there is a virtually imperceptible and seemingly unimportant detail that in Pushkin distinguishes between a siren and a rusalka. And it is precisely this detail that is our entry into the semantics of rusalka's representation.

---

1   On Pushkin's "Rusalka", see Emerson (2019), Wachtel (2011), and Zaharov, Lukov and Lukov (2015).

Namely, sirens are traditionally *daytime* creatures, because desire is the strongest then; in medieval scriptures this libidinal surge was called a "noonday demon", for it appears when the sun is high and the heat makes a monk lazy and inert (Agamben 1993: 3–10). This trait, first seen in Homer, inconspicuously persists throughout the Victorian and early modern siren literature.[2] Rusalki are, on the other hand, pale *midnight* demons (like in Orest Somov's "Rusalka" ([1829] 2016)) that almost exclusively appear at night, when "all is still" (Gogol [1831] 1957: 97) and the landscape is bathed in moonlight – paleness of rusalka's body reflects paleness of the moon. This trait is so characteristic of the nineteenth-century rusalka poems and paintings that Pruszkowski picked it up and pushed it to its limits. The moon in his painting is young, cutting the light source of the scene and turning its background into shadows and contours. And yet, light shines from its centre, contrasting the dark thicket – it shines *from among* rusalki. Their faience skin produces the moonlight of its own, becoming the source of light and illuminating the scene from within.[3] What we learn from Pruszkowski is that even when the moon does not allow it, the connection between rusalki's paleness and moonlight is indissoluble. As Nikolai Gumilev ([1905] 1998: 62) said, "The rusalka has a twinkling gaze/The dying gaze of midnight."

This paleness/moonlight trope is found in most rusalki narratives.[4] Gogol's Levko from "May Night, or the Drowned Maiden" sees a rusalka "white all over, like a sheet, like the moonlight" (Gogol [1831] 1957: 98); Turgenev's Kostya from "Byezhin Prairie" recounts that as "the moon was shining bright [...]" the rusalka was "herself as bright and as white" (Turgenev [1852] 2014: 99).

---

2   We see daytime mermaids in F. Anstey *The Siren* (1884), Hanry Carrington's *The Siren* (1898), Bret Hart's *The Mermaid of Lighthouse Point* (1901), H. G. Wells' *The Sea Lady: A Tissue of Moonshine* (1902), E. M. Forster's "The Story of the Siren" (1920), and especially in the summer heat of Guissepe Tomasi di Lampedusa's "The Professor and the Siren" (1961); but these are only highlights. On late Victorian and early modern siren literature, see Teodorski (2021).

3   The same play of contrasts can be seen in Konstantin Makovsky's 1879 *Rusalki*.

4   The only examples, to my knowledge, that fuse rusalki and the sun are Lermontov's "Rusalka" (1832), where they kiss the drowned knight at noon, and Gogol's short story "Viy" (1835), where the philosopher Khoma Brut sees a rusalka in the water surface followed by the sun. But even here, the sun is only a reflection/reversal of the night when "the waning crescent of the moon was shining in the sky" (Gogol [1835] 1958: 198).

Slavic Myth

There are examples, however, that articulate this trope in broader terms: in Pushkin's drama *Rusalka*, this trope metastasises over the narrative as a whole. The plot draws on a number of sources[5] and tells of a peasant girl left pregnant by the local prince. In an act of desperation, she jumps into the Dnepr and turns into the Queen of rusalki. Years later, she sends her child to lure the prince and drown him in the Dnepr. Unfortunately, this is where the story ends: Pushkin left it unfinished.[6] What is important for us here, though, is the fact that *Rusalka* can be divided into two parts – before and after the turning – the transition that have been interpreted as Pushkin's artistic collision of realism and fantasy, literature and folk mythology (Zgurskaja 2014). But here I would like to point that the first part happens during the *day*, while as soon as the girl turns in rusalka the *night* falls, and everything is dipped in moonlight. "Sisters, leave your spinning", says Queen Rusalka at the beginning of the second part. "The sun has set. Enough. A shaft of moonlight gleams above us" (Pushkin [1837] 1982: 236). Seventy years later, in 1910, when Vasili Goncharov made a silent movie based on Pushkin's play, this trope was still alive: as the Price is lying in the Queen's lap, dark background contrasts everything else (shells, corals and rusalki themselves), simulating a moonlit scene at the bottom of the Dnepr.

Rusalki and moonlight are inseparable, creating one half of the quadrangular semantic network through which Mikhail Lermontov spoke: "The rusalka floated on the blue river, / Illuminated by the full moon; / And she tried to splash to the moon / Silvery foam waves" (Lermontov [1932] 1896: 58).

5   Apart from Motte Fouqué's *Undine* that Vasily Zhukovsky translated into Russian in 1830s, one of the big precursors to Pushkin's "Rusalka" was Nikolai Krasnopol'skii's four-part musical *Dneprovskaia Rusalka* (1803–7). However, this play was a russification of two Viennese *Singspiele*, Ferdinand Kauer's *Das Donauweibchen* (1798) and Karl Friedrich Hensler's *Die Nymphe der Donau* (1803) (Emerson 2019: 177).

6   Throughout the nineteenth and twentieth centuries, there were numerous attempts at finishing Pushkin's *Rusalka*, from Alexander Dargomyzhsky's opera *Rusalka* (1856) to Vladimir Nabokov's "Rusalka" (1942). It is interesting to note that more than a hundred years after Pushkin, the moonlight remains the background for the figure of rusalka. "We call the moon to the river wedding", sing rusalki in Nabokov's version, and as the Prince drowns a ray of moonlight falls on his forehead (Nabokov-Sirin 1942: 184).

In 1871, at the first Itinerant exhibition, Nikolai Kramskoi presented himself with the central piece called *Rusalki*.[7] The painting was the illustration of Gogol's "May Night, or the Drowned Maiden", and Kramskoi envisaged it as a study of moonlight, Ukrainian landscape and folklore.[8] It depicts a group of rusalki lounging at the lake's shore. They are pale, drenched in moonlight, clad in white robes that resemble night gowns. Some of them prowl in the reed, some are bathing, but the one in the centre-right has her eyes on the absent moon, her stargazing emphasised by her clasped hands. The trope of paleness/moonlight is obvious here,[9] but the scene exudes stillness and reverie as if it were a dream.[10]

Kramskoi's painting emphasises the second part of the trait network in the representation of rusalki, and that part heavily relies on folklore: it is said that sleeping on the ground during the *rusalnaya nedelya* can make one mad or fatally ill (Zečević 1974: 110). This trope is widely present in literature and painting, turning the encounter with rusalki into a hallucination or a dream. The dream does not, however, come alone, the same as paleness follows moonlight: when the protagonist (the victim) falls asleep and is not sure whether he dreamed the encounter with rusalki or not, the surrounding night falls asleep with him and becomes perfectly *still*. "Did he see this or did he not?" asks Khoma Brut from Gogol's "Viy", "Was he awake or dreaming?" (Gogol

7   In-depth studies of this painting are few, but see Štejner (2006).
8   Kramskoi actually travelled to Ukraine to study the local landscape (Karpova 2000: 10; Kononenko 2009: 13; Stasova 1887: 18).
9   *Rusalki* belongs to a group of few Kramskoi's paintings (*Somnambula* (1871), *Moonlit Night* (1880)) that directly study moonlight. The painting must have been more than just a study, since Kramskoi considered the moon as a source of enlightenment. "What good is the moon, this plate?" says Kramskoi (1954: 188). "The flickering of nature under these rays is a whole symphony, powerful, high, tuning me, the poor ant, to a high spiritual order."
10   Critics generally agree that the main characteristics of this Kramskoi's study is the pervading silence and dreamlike state, created by the intersection of the realistic (landscape, girls) and the mythological (rusalki) (Brunson 2016: 175–6; Karpova 2000: 10; Kononenko 2009: 13; Kuročkina 1989: 28; Orlova 2014: 13). Some even compare it to his other 1871 work *Somnambula* (Orlova 2014: 13), and quite astutely, because here the play of moonlight, paleness, stillness and dreaming is overcharged by the eponymous narrative of sleepwalking.

[1835] 1958: 199). Khoma Brut gallops through the night with a witch on his back, while "[t]he forests, the meadows, the sky, the dales all seem as though slumbering with open eyes; not a breeze fluttered anywhere" (Gogol [1835] 1958: 198).

Sleep/reverie and stillness depend on each other, complement, define, consume and appropriate each other. The hero's slumbering body is one with the night's stillness the same as rusalka's body is one with the moon. As Gogol's Levko was "getting numb" and his "head drooped" under the moonlight, "[t]he motionless pond sent a breath of refreshing coolness at the tired wanderer and lured him to rest for a while on the bank. All was still" (Gogol [1831] 1957: 97). The same motionless pond will soon show him the image of a rusalka "white all over, like a sheet, like a moonlight", and it is precisely in this image, in this interconnectedness of paleness, moonlight, sleep/reverie and stillness that the singularity of the rusalka figure emerges, setting it apart from all other water maidens. In the centre of them all, like the first among many, rests still-ness, the ultimate gift and power of rusalki. Stillness of nature, stillness of the body, stillness of life, stillness of desire: stillness as the ultimate aesthetic of *drowning*. Sinking to the bottom with her, the male victim becomes "numb", inert – one could say, in folklore terms, fatally ill. However, this illness is the rusalka's gift, the gift of being drowned by the drowned; there is a semantic exhaustion at play here in which the drowned reduces another to itself and the signifier completely overlaps with the signified – no leftovers, no traces.

What rusalki have to offer is the final, semantic and libidinal stillness.

Sirens are libidinally irresistible; they lure men with promises of ecstasy and pray on them by the semantically differential nature of that lure (as well as of their half-human bodies). Rusalki, on the other hand, are rarely irresistible (sometimes they are even plainly ugly[11]), but they force upon their victims the resolution of deep internal conflicts and abandonment of an agonising pain. In Dargomyzhsky's 1856 opera adaptation of Pushkin's *Rusalka*, the Prince, torn by gilt, jumps into the river and finds peace; he becomes libidinally and semantically still. This resolution is the operatic precursor to Goncharov's last

---

11   "[L]ike a dace or a roach, or like some little carp so white and silvery" (Turgenev [1852] 2014: 99).

scene: the prince is prone in the Rusalka's lap, while the scene enacts a tableau that turns living beings into statues, objects.

Meanwhile, the rusalki are looking at the prince wondering in the words of Lermontov:

> And there on a pillow of bright sands
> Under the shade of dense reeds
> The knight is sleeping, the prey of the jealous wave,
> There sleeps a foreign knight.
> [...]
>
> But to passionate kissing, not knowing why,
> He remains cold and dumb;
> He sleeps, and as I hold him against my breast,
> He doesn't breathe, or break his sleep with even a whisper!
> (Lermontov [1932] 1896: 58)

What rusalki give to their victims is a painless peace of inanimation; the cancellation of their ceaseless differentiality; the reduction to sameness, the ultimate utopian peace.

Leila Anani

Tropic of the Sea (Satoshi Kon, 2013)

The mermaid in mainstream manga and anime generally draws more from Western ideals than traditional Japanese folklore. Toei Animation famously adapted Hans Christian Andersen's *The Little Mermaid* in 1975 and many of the subsequent depictions of mermaids in anime and manga have copied this romantic image.

It features a cute, young female with long hair (often blonde), BESM (big eyes, small mouth), a musical voice and a fluid fish tail with fluted fin. *Mermaid Melody Pichi Pichi Pitch* (Hannamori and Yokote, 2004), *Seto no Hanayome* [My Bride is a Mermaid] (Kimura, 2007), *Bamyūda Toraianguru: Karafuru Pasutorāre* [Bermuda Triangle: Colorful Pastorale] (Nishimura, 2019) and Princess Shirahoshi from *One Piece: Wan pīsu* [One Piece] (Oda, 1997–present) are all examples of the cute Western style mermaid in manga and anime. The majority of mermen follow the same template with cute-looking men with fish tails as seen in *Orenchi no Fura Jijou* [Merman in My Tub] (Itokichi, 2014) and *Mermaid Boys* (Serachi and Yomi, 2018), among others.

However, there is more to the mermaid in anime and manga than this kind of superficial, romantic beauty. Reiko Shimizu's fascinating shojo sci-fi manga series *Moon Child* (1989–93), Satoshi Kon's manga *Tropic of the Sea* (2013), and Masaaki Yuasa's recent anime film *Yoake tsugeru Rû no uta* [*Lu over the Wall*] (2017) all look at environmental themes through the lens of mermaid myths.

It is this facet of the mermaid, as nature's emissary that this essay will focus on, looking at the three main functions of the merperson in this role – prophet, avenger, and/or saviour. Subsequently, this essay will also address why the mermaid, above all other mythical creatures, is most suited to undertake these functions. The focus for this will be Kon's *Tropic of the Sea* due to the nature of its Japanese setting, which gives it a unique cultural and historical

Figure 13. Iconic anime mermaid image from *Anderusen dōwa ningyo-himi* [The Little Mermaid], directed by Tokoharu Katsumata and Tim Reid (Toei Animation, 1975).

perspective as well as exemplifying all three of the aspects mentioned above in relation to environmentalism.

The story of *Tropic of the Sea* is superficially quite simple: Yosuke is the latest descendant in a family of shrine guardians who receive a mermaid's egg once every sixty years which they have to protect until the mermaid returns to reclaim it. In return for their service, the mermaid blesses the coastal town with calm seas and plentiful catches of fish. However, times change, and the younger generation no longer believe in tradition or legends of merfolk. Yosuke's father, lured by greed, sells the temple land to a wealthy property developer who wants to turn the egg-shrine into a tourist trap. Once he finds out about its magical healing properties, he wants to exploit it and refuses to return it to the sea. This incurs the wrath of the mermaid whose vengeance threatens to destroy the whole town.

Unlike many every other manga and anime featuring mermaids, *Tropic of the Sea* is firmly set in Japan and isn't influenced by Western traditions to try and widen its audience and appeal. As such it presents a specifically Japanese mermaid which requires closer consideration.

Modern Japanese Myth

Japanese Mermaids

Western ideas of Japan often evoke two opposing images: first, Tokyo as the technological capital of the world as a vision of skyscrapers, overcrowding, the foremost producer of electronics, one of the biggest importers of fossil fuels with a huge nuclear waste problem; second, the historical (pre-Meiji period) Japan, which was a rural island culture, steeped in tradition and myth, dependent on fishing and the bounty of the sea. Both are equally represented in film, manga and anime, though usually kept very separate, except on one major topic: environmentalism. Godzilla is the iconic example; an almost mythic monster rising from the sea to destroy Tokyo – showing nature's violent revenge on man's arrogance. Godzilla though, having been mutated by nuclear testing, can still be seen as a man-made product of the industrial age. Satoshi Kon cleverly fuses these two versions of Japan far more effectively in *Tropic of the Sea*, where he essentially uses the myth of the mermaid (the past) to oppose an urban development plan (the future).

The Ningyo (man-fish) is the nearest equivalent of the mermaid in traditional Japanese folklore but is subtly different from its Western counterpart. Usually male, he is a hideous, small creature with a fish's body and humanoid head and arms. There is no romance or seduction associated with the Ningyo, and marine sexual symbolism in Japanese culture is usually connected to the octopus, most famously in Hokusai's shunga "Dream of the Fisherman's Wife" (1814), in which a woman is being pleasured by a giant cephalopod. Other than his appearance as a monster *du jour* in various Yokai-populated manga series such as Kitaro, the Ningyo features surprisingly seldom in manga and anime.

Tropic of the Sea is of interest as, just like its fusion of two opposing concepts of Japan, it also fuses the Ningyo and the mermaid. In appearance the mermaid is Western: half-woman, half-fish, with long fair hair. However, she is neither cute nor sexualised. Her seaweed-like hair modestly conceals her breasts, her fingers are webbed and she is much larger than our hero Yosuke. She is the mother goddess, who gifts her egg to a mermaid temple to be tended by priests for the sixty years it takes to hatch.

The mermaid temple from the story comes directly from the mythology of the Ningyo. A number of mummified mermaids (resembling the infamous

Figure 14. Yotsuke and the Mermaid. *Tropic of the Sea*, by Satoshi Kon (Kodansha: 2013).

Fiji mermaid) are still held in temples today. Indeed, the temple in *Tropic of the Sea* bears some similarity to the Tenshou-Kyousha Shrine in Fujinomiya, which holds the largest and oldest mermaid mummy in Japan. The Fujinomiya Ningyo has its own legend attached to it, where it appeared to Prince Shotoku at Lake Biwa and with its dying breath told its tragic tale. The Ningyo had been a fisherman in life, who, wanting a bigger catch, had trespassed into protected waters and was transformed as punishment for his greed. During his years as a Ningyo he had learnt his lesson and come to understand the horror of killing unnecessarily. He was now ready to move on to the next life, but begged the prince to display his remains in a temple to remind people of the sanctity of life.

While *Tropic of the Sea* does not reference this legend specifically, the message is the same and in both cases the mermaid is used as a vehicle to warn people about the evils of greed and to promote harmony with nature. This idea of the mermaid acting as an environmental emissary leads directly into the three ecological aspects of the mermaid mentioned at the beginning of this essay, and the first of these being the mermaid as prophet.

In its role as a prophet the mermaid is an almost omniscient, wise and caring, godlike being coming from outside the human realm to warn mankind and offer a chance at salvation with a divine "gift". The mermaid in *Tropic of the Sea* follows this mould as she comes from the unexplored depths "way down deep in the ocean where no one can see", which is as effectively beyond the human world and as unknowable as heaven. The mermaid's egg (the mermaid's one and only daughter) is entrusted to the human world and can perform miracles, consciously nodding to Christianity's idea of the Son of God. It also falls into the wrong hands as the greedy land developer Kenji Ozaki takes the egg and thinks he can exploit its power. "It is an asset that will pay inestimable dividends to mankind /Give it back to the sea?! Those idiots have no idea how valuable it is!!" (Kon 2013). While Kon's mermaid does not offer eternal life, there are famous Japanese Ningyo legends such as "the 800-year-old nun", where eating the flesh of the mermaid grants extended life or immortality. This forms the basis for Rumiko Takahashi's manga *Mermaid Saga* (1984–94) and it's anime incarnations *Ningyo no mori* [Mermaid Forest] (Mizutami, 1991) and *Ningyo no kizu* [Mermaid Scar] (Asaka, 1993).

While the mermaid in *Tropic of the Sea* foretells the consequences when humans exploit the environment for personal gain, it should not be forgotten that it is the mermaid herself who will be undertaking those consequences. This then brings us to the second category on the list: the mermaid as an avenger.

As stressed in *Tropic of the Sea*, if nature is treated well, it rewards you – "in return it is said we have been blessed with calm seas and bountiful catches of fish" – but abuse it and you suffer divine punishment: "They're going to destroy the island and fill in the beach/and we can't catch a single fish!" (Kon 2013). Indeed, the mermaid is so upset with Ozaki's plan to keep the egg that the weather is cloudy and "The sea is dead" (Kon 2013). This resonates with another of the myths about the Ningyo where they are harbingers of storms. If a fisherman were to accidentally catch one, he should immediately throw it back into the sea or suffer a terrible storm. An illustration of this is seen in volume 1 of Toriyama Sekien's *Konjaku Hyakki Shūi* (1781) – a kind of Japanese bestiary – which has an image of a Ningyo at the centre of a great wave. This is very reminiscent of Hokusai's later *The Great Wave* (c. 1829), minus the mermaid, which is a classic representation of nature's might. In line with this, in *Tropic of the Sea* the mermaid creates a tsunami to obliterate Mr Ozaki's resort development, with the realisation that "the sea is angry that we broke our promise" (Kon 2013).

However, the mermaid does not just destroy, she also heals enacting the last category on the list, in becoming a saviour. In *Tropic of the Sea*, personal salvation is both physical and spiritual. As well as healing a cut and curing cancer, the mermaid rescues Yosuke from drowning twice, ensuring his personal safety. She also rekindles his spirituality, replaces his sense of wonder and heals his soul. Jaded by the death of his mother and his emotionally closed-off father, Yosuke has ceased to believe in mermaids: "Is that egg even real?/ course not" (Kon 2013). Yet by the end of the story, just like the cursed fisherman in the Ningyo legend mentioned above, he has gained enlightenment and realises the importance of the egg: "That egg is indeed a treasure given by the sea/it wasn't a gift! It's a precious item in our charge" (Kon 2013). Mr Ozaki too learns his lesson and thanks Yosuke, scaling down his development and enlisting the whole city's co-operation, whereas previously he'd only had "slaves to the dough of real estate agents" (Kon 2013).

The relationship between the mermaid and environmentalism has so far been clear-cut in the narrative of *Tropic by the Sea*: progress is bad and the mermaid as nature's emissary wrecks a terrible vengeance. Yet, it's not quite that simple and to see how we must look again at salvation, but on a larger scale. While set in modern Japan, the location of *Tropic of the Sea* is Ade, a small coastal Japanese town, steeped in tradition, existing "thanks to the sea's bounty" (Kon 2013). However, the town is slowly stagnating, in exactly the same way as Hinashi in *Lu over the Wall* (2017), mentioned earlier. While the older generation (grandpa) clings to the old ways, the next generation are struggling to make a living and the youth are moving away to the cities. Nami has only returned to the village due to a failed relationship and you get the feeling that Yosuke does not really want to be the next priest, just like Kunio, the main protagonist, in *Lu over the Wall*. Progress in and of itself is not a bad thing, as noted in *Tropic of the Sea*, "Mood and nature are important, but people can't live on pretty scenery alone" (Kon 2013). You can clearly see both sides of the argument. It is only when you add greed into the mix, in the figure of Mr Ozaki, that it becomes rotten "Like a whore who got lots of surgery and wears heavy make-up" (Kon 2013) with those who have sold their souls to greed becoming pimps.

The mermaid comes along, encounters human hostility at the hands of greedy developers, but finally after the goodness of the hero shines through, saves the town. In *Tropic of the Sea* the mermaid sends a tsunami, while in *Lu over the Wall* there is a flood although it is initiated not by the mermaids themselves, but rather by a curse laid on the town years ago when they killed a mermaid in fear. However, its effect is just as cleansing. In both cases the slate is wiped clean, enabling the town to have a fresh start. In both narratives, the mermaids disappear once the flood has abated and performed its cathartic function, yet each leaves it open to the possibility of their return. Yosuke asks in the final scene of *Tropic of the Sea*, "Does this mean our pact with the mermaids is over?", to which his grandfather replies "Who knows?" (Kon 2013).

By the end of *Tropic of the Sea* and *Lu over the Wall*, the future is bright, and harmony has been restored. The conflict the Japanese setting gave rise to tradition vs progress and mirrored the wider environmental conflict of nature vs man. The mermaid was the instrument in bringing about change and restoring balance. The sea and land are no longer at odds, there is a middle ground between nature and progress, and the characters have revived their spirituality. They have also come to terms with the loss of loved ones; Yosuke finally accepts the explanation of his mother's death, and in *Lu over the Wall* Kai comes to terms with his mother's abandonment as well as various other characters (Granny Octopus and Grandpa) being reunited with their transformed loved ones.

Harmony is the key word when looking at why the mermaid is the perfect emissary for mother nature. It suggests balance on the widest scale: nature and mankind working together. It also suggests people working together rather than being at odds. In *Tropic of the Sea* the notion of harmony is symbolised visually by the mermaid egg. In one image we see Yosuke holding the orb and it deliberately resembles the globe – with a mystical, external field (water?) taking the shape of the continents/oceans. It's a powerful image – he's flanked by his two friends – and so we get all multiple notions of harmony in the one frame. It also suggests that the power is held by the youth of today. There's a further image of the egg with Yosuke alone holding the giant orb with a baby mermaid curled up inside it, which in many respects summarises the entire story: Yosuke's hand is visible through the translucent shell with the power to cradle or kill the foetus within. It reminds us of the fragile balance between mankind and nature, that could be shattered at any moment, yet the image of the foetus is also one of hope.

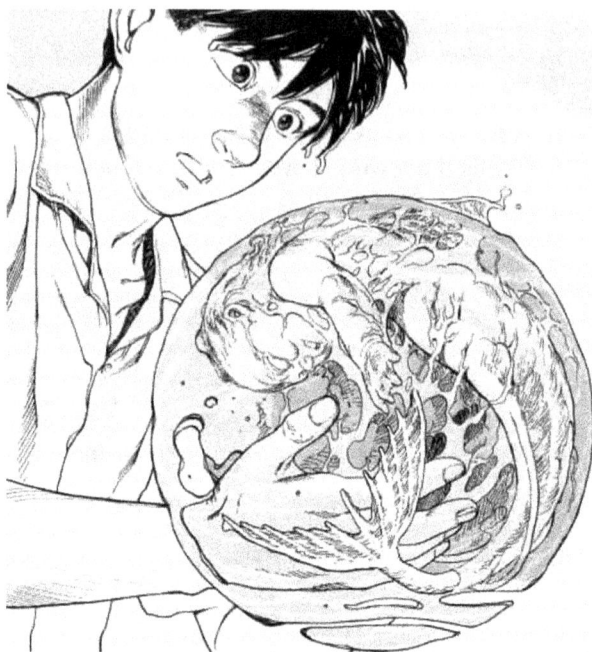

Figure 15. He's got the whole world in his hands. *Tropic of the Sea*, by Satoshi Kon (Kodansha: 2013).

Part II

Femininities and The Deep

Philip Hayward

Duyung (Abdul Razak Mohaideen, 2008)

*Duyung*s and Mermaids in Malay Popular Culture

There are various folkloric and literary accounts of riverine and coast-dwelling female aquatic humanoids around the Malay peninsula and the present-day states of the Malaysian federation on the north coast of Borneo.[1] Terminology is complex here, as the aquatic mammal known in the West as the dugong (species *dugong dugon*) is referred to in Malay as the *duyung* (occasionally *duyong*). The same term is also used to refer to aquatic, tailed humanoids, sometimes with the term *putri* (daughter) as a prefix to indicate gender and human aspects, as well as with the term *ikan duyung* ("fish" *duyung*) also being used to distinguish the humanoid from the sea mammal. Despite the close association between the sea mammal and aquatic humanoids, contemporary popular cultural representations of *duyung* have rendered them in forms closely resembling western mermaids. Drawing on shifts in the representation of Japanese marine folklore in popular culture, Hayward (2018a) has described this process as a progressive *mermaidisation* that is akin to an invasive species taking over a biological niche and becoming part of a local environment. Similar processes have been evident in Malaysia and have resulted in the media texts discussed below.

While mostly associated with coastlines, modern Malaysian *duyung* folklore derives beliefs of communities such as the proto-Malay Temuan people of the south-western Malayan peninsula. This (largely inland) community has

---

1   Thanks to Alistair Welch and Sharina Abdul Harim for research assistance.

a number of beliefs about dangerous *duyungs* threatening those who venture into particular rivers or lakes and also includes a belief that *duyungs* are transformed human females and that their tears have magical properties (Teip, Baer and Mohamad 2016: 36). Accounts of *duyungs* in areas of the Malaysian coast frequented by dugong are also common and have been circulated in the "media-lore" of the Internet, press and broadcast media. One notable example concerns two adjacent, island-like peninsulas in the state of Terengganu named Pulau Duyung Besar and Pulau Duyung Kecil. Local folklore attributes the names to two *duyungs* that once washed ashore in the area. In online renditions of this tale, such as Rush's (2017), the *duyung* and the islands are referred to by the English-language term "mermaid," modifying the original associations of the place names in a manner that reflects international culture and enhances the area's tourist appeal.

One of the first, and the best known, literary accounts of aquatic humanoids in Malay culture occurs in the sixteenth-century compilation of genealogical tales originally entitled *Sulalatus Salatin* (often referred to as the *Malay Annals*). In these, the Malay noble, Raja Chulan, is identified as having married a princess, Mahtabul Bahri – the daughter of an undersea king named Fatabul-Ard. While English-language versions and reiterations of this narrative characterise the daughter as a mermaid, she is not referred to as a *duyung* in the original and there are no grounds for assuming that she was understood to be fishtailed, like modern Malay *duyungs*. Indeed, Hussain Othman (2008: 98) has argued that the aquatic realm can be understood as a representation of "the underworld, the world of the dead" more generally, which Raja Chulan visits and assimilates into the terrestrial rather than as the realm of merfolk.

The first modern popular cultural representation of *duyung* folklore occurred in 1964, in the form of M. Amin's film *Ayer Mata Duyong* ("The Duyong's Tears") – made well before the western wave of mermaid-themed cinema that followed *Splash* (1984) and *The Little Mermaid* (1989). While the film draws on folkloric themes, it is based around a fictional story of doomed love between a fisherman, named Awang Jermal, and a princess named Tengku Intan. After growing up together in a coastal village they become reacquainted as adults when Awang becomes a bodyguard at the royal court that the princess frequents. Their proximity stirs strong affections that cause jealousies and prompt a courtier to remind the Sultan that an earlier romance between his great-grandmother and a commoner was accursed, resulting in her being transformed into a *duyung* and turned later into stone. After an unsuccessful attempt by courtiers to have Awang sent away and killed, he returns and his marriage to the princess is approved by the Sultan. However, fate intervenes, and the dreaded curse is again enacted on the day of their wedding. During a ritual bathing ceremony, a storm blows up at the beach and both Awang and the Princess are swept away into the waters. Awang manages to regain the shore but pines and sickens as his bride-to-be does not return. His stepsister, who has held affections for him for many years, then takes him back to their village (implicitly to develop a relationship). Witnessing this, the princess, who has survived but has become transformed into a *duyung*, swims away, never to be seen again.

The *duyungs* featured in the narrative – that is, the Sultan's great-grandmother, only seen frozen in stone, and the princess, transformed on her wedding day – appear as cursed, tragic and degraded/deformed women. The film essentially delivers a cautionary tale, relating the perils of marrying across the class divides that exist between nobles and commoners. In this regard, the film manifests a conservative sensibility articulated in the year after Malaysia gained its independence and at a time when Malay society was shifting towards a more egalitarian and democratic focus. In this manner, the film can be considered as nostalgic for the fading days of aristocratic power and rigid class structures. The *duyung* symbolises the physical distortion perceived to result from disruption in established systems of social organisation. *Duyungs* are represented onscreen three times in the film. The first occurs during an animated title sequence where a long-haired *duyung* is rendered with fish-scales extending to her upper torso while leaning on a hand-drawn rendition of the film's title. The second appearance is as a briefly glimpsed stone transformation of the Sultan's great-grandmother, kept behind curtains in a dusty locked room. A living *duyung*, in the form of the transformed princess, only appears in the film's final scenes, swimming and, in the image that gives the film its title, weeping as she rests on a rock as her intended departs. As this summary suggests, the mermaid is not posited as a creature in her own right but rather as a transformed human and there is, consequently, no representation of a broader "mer-realm" with particular *duyung* attributes.

Amin's representation of the *duyung* was in marked contrast to the next significant rendition in Malay popular culture, produced by artist and designer A. Ghafar Bahari in his *komik* publication *Puteri Duyung* (Suharto, 1985), which chronicled a human male's attraction to a captivating *duyung*. The *komik* was typical of a wave of Malaysian and Indonesian publications in the eighties that drew on and dramatically rendered folkloric themes (see Hayward 2018b: 98–9). As the cover of Bahari's *komik* makes apparent (Figure 1), his *duyung* contrasted to the more demure, transformed princess in Amin's film by being a powerful, erotic and somewhat threatening figure. Her long tresses stream away in the water, seashells cover the nipples on her large breasts and her fish scales spread from the tip of her tail to just under her cleavage. Her expression is far from friendly, and she has less in common with the *duyung* in Amin's film (or the young adolescent mermaid of Hans Christian Andersen's seminal "The Little Mermaid" (1837)) than she does with renditions of the *naga* snake-women featured in Indonesian popular culture at the time (Hayward 1997: 94–6).

Abdul Razak Mohaideen's 2008 feature film *Duyung* (*Mermaid*) differs from the preceding texts by combining elements of Malay folklore with an environmental theme. The lead male character, Jimmy, a self-appointed environmental guardian for his local coast, collects rubbish and abstains from eating fish. He is supported by an unlikely ally, an ape-man of a species identified as an "Orix" in the film.[2] Jimmy is enthralled with stories about a local *duyung*, believing that he saw her as a child, and his mother prompts him to look after the environment by saying that the *duyung* will only return if the waters are kept clean. Jimmy is also smitten with a local girl named Aspalela but her father, who is also the village leader, requires any suitor to have his own house, which Jimmy does not as his family does not own land. His unusual solution is to construct a floating home from rubbish that he has collected, which is promptly dismantled by a rival suitor, named Kordi. At this point Jimmy finds a *duyung* in a fishing net and releases her. The grateful *duyung* gives Jimmy a seashell to blow into if he ever needs her. But despite her specification that he should tell no one else about her, he blows the shell when he is with his fellow villagers to prove to them that the *duyung* exists. Fortunately, she

---

2   Orix resemble the Orang Mawas, a hairy ape-man of Johor folklore.

Figure 16. Comic of Putri Duyung commonly associated with the 1985 film of the
same name by Atok Suharto.

realises and does not respond. Forgiving him, she takes Jimmy to her marine
lair and learns about his love for his mother, the Orix and Aspalela. While
she is sad that Jimmy does not love her in the same way, she allows him to
return to shore and leaves a magical gift that makes him prosperous and boosts
his chances of becoming a successful suitor for Aspalela. Kordi responds by
going to Jimmy's floating home, sounding the shell and capturing the *duyung*,
intending to sell her to a local funfair owner. But all ends happily for Jimmy

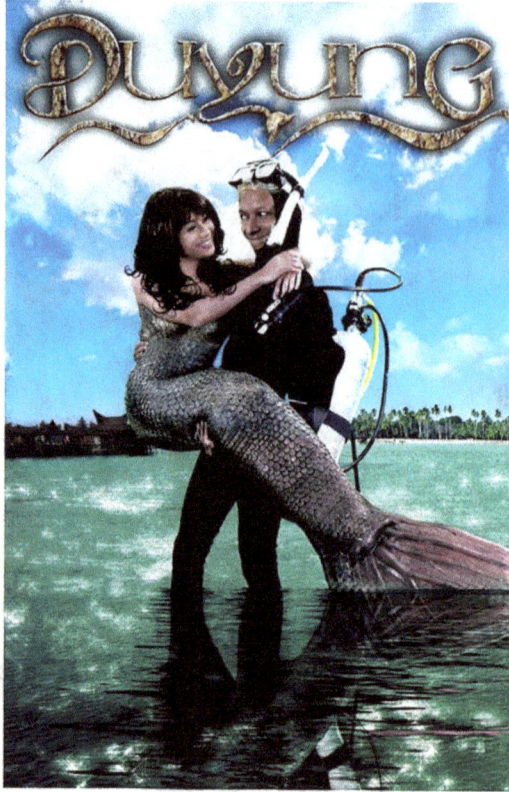

Figure 17. Promotional poster for *Duyung*, directed by A. Razak Mohaideen (KRU Studios, 2008).

as he returns, releases her, marries Aspalela, opens a successful eco-tourism business and raises a family.

Like her cinematic predecessor, the *duyung* in Mohaideen's film, *Duyung* (2008), is shown as covered in fish scales that stop below her collar area and she has long, dark tresses. In these regards she is far more modest than Hollywood mermaids such as Madison from *Splash* (1984) or the female protagonist of the *Puteri Duyung* comic, as befits a character in a film made for family audiences in in a predominantly Muslim nation. But unlike the character in the 1964 film, she is unequivocally a *duyung* (rather than a woman transformed into one). While she shows feelings for Jimmy, she is effectively used as a plot device that enables Jimmy to gain success in his eco-tourist venture and can, in this regard, be perceived as a spirit-of-the-sea. Together with the inclusion of the Orix, this eco-tourism theme is the most novel aspect of the narrative. Mermaids have been associated with various international environmental protests and advocacy groups in recent years (see, for instance, Bennett 2015 and Priestley 2013) but the association of *duyung* with environmentalism was innovative in Malaysian contexts.

The most recent Malaysian audio-visual production to represent fishtailed humans is a 24-episode fantasy series entitled *Duyong Aridinata* broadcast in 2010. The series depicts the adventures of a group of aristocratic, fishtailed, merpeople living under the sea in a mirror of human terrestrial existence. Its scenario reflects aspects of Hans Christian Andersen's "The Little Mermaid" and its re-imagination by Disney in Clements and Musker's 1989 film adaptation. The film's lead character, Prince Aridinata, earns the displeasure of the king of the undersea realm by refusing to marry the mer-princess Antartinka and leaves the sea to seek his fortune on land, with his tail being transformed into legs by a female, submarine shaman. As well as echoing western tales, this scenario reverses the plot device of *Ayer Mata Duyong*, in which *duyungs* are cursed and transformed humans. The scope and dramatic complexity of the series, and its extended underwater sequences, mirror similar Indonesian *sinetron* TV dramas such as *Putryi Dyung* (Suharto, 2001) and *Sissy Si Putri Dyung* (MD Entertainment, 2007) and can be seen as more representative of an early twenty-first-century pan-Southeast Asian sensibility than the earlier, more overtly Malay productions discussed above. In this manner, the *duyung* legends that inspired twentieth- and early twenty-first-century Malay popular culture are fading and being replaced by the increasingly popular, western-derived figure of the mermaid, who has her own set of associations that ultimately derive from Disney and Andersen rather than peninsular folklore.

The success of *Duyung* in Malaysia and growing publicity over international mermaid performers (Henault 2020) spurred Malaysian model Felixia Yeap to become Malaysia's first professional mermaid performer in 2013, inspired by western mermaid cartoons such as *The Little Mermaid* and by professional mermaid models in Australia and the United States (Yee Yun, 2013). Her venture was regarded as controversial due to the skimpy nature of her upper bodily costume in publicity photos, with her breasts covered by two fabric shells. Unfavourable responses to media coverage of her career move contributed to her quitting mermaiding and modelling soon after and converting to Islam, adopting the name Raisyyah Rania Yeap. Six years later the cultural climate was more favourable to such ventures, with Malaysia's first mermaid training school opening in the coastal town of Klang (No Comment TV 2019) and with promotional images showing its students dressed similarly to Yeap, seemingly without controversy. The international orientation of this school's "mermaidery" was emphasised in a promotional video in which footage of training sessions was set to the song "Mermaid Party" from the animated US feature *Barbie: The Pearl Princess* (Zeke Norton, 2014).[3] This suggests that in the light of recent popular cultural developments, Malay *duyungs* are increasingly local manifestations of transnational mermaids rather than modern renditions of the type of traditional figures formerly associated with the term.

3   Online at: http://www.dolphinlee.com/mermaid/.

Laura Sedgwick

The Legend of Kópakonan (1891)

This essay will consider the figure of the female selkie as an example of the "monstrous feminine" through the lens of the Kópakonan legend from the Faroe Islands. It will also explore the shape-shifting nature of the selkie and discuss why such an apparently harmless supernatural creature would receive vicious punishment.

The coastline is essential to tales of the selkie, where the coast represents a border that offers "both a supernatural threat and defense" (Harris 2009: 5). Within selkie legends, it is across such borders between the wilderness and cultivated land that the supernatural intrudes. The status of the coast as "a dangerous frontier" becomes clear when we consider the historical context of the nations around the North Sea (Harris 2009: 6). Yet in the tale of the selkie, it is not the Viking invader that crosses the land/sea boundary, but rather a gentle creature that seemingly poses little threat to the humans nearby.

The selkie legend appears along the coasts of Ireland and Scotland, particularly clustered around the Orkney Islands, and the tales stretch to the Faroe Islands and Iceland. At the most fundamental level, selkies are shapeshifters. They take the form of seals in the sea and shed their seal skin to become humans on land. They may only willingly stay around humans for a short time. Once they don their seal skin again, they cannot return to shore for another seven years. In some stories, a human man wakes one day to find his wife gone. It is only with her departure that he finally understands that he married a selkie. There is some consternation as to the difference between "Finfolk," human magicians capable of taking seal form, and the "Seal Folk," as discussed by Alan Bruford (1997). For Sigurd Towrie (2021), the "original darker, malicious nature" of the selkies in the Orkney Islands was lost, though the fragmentation between the Finfolk and the selkie-folk tales may have been a faulty interpretation of categorisation by nineteenth-century folklorists. Yet it is the legend of the selkie, or "seal bride" that will occupy our attention here.

The "seal bride" is created when humans force selkies to stay on land by stealing their seal skin. Reidar Thoralf Christiansen (1977) assigned the "seal bride" story the "Migratory Legend" – (ML) code 4080 – though the stories are more numerous along the Atlantic coasts of Iceland, Ireland and Scotland than they are in Norway. In the Orkney story of Goodman o' Wastness, a bachelor steals a selkie skin and forces the selkie to become his wife (Towrie 2021). She bears him seven children but eventually her youngest daughter tells her where the Goodman keeps the seal skin. The selkie wife regains her seal form and returns to the sea and her selkie husband. The Goodman spends his later years wandering along the shore, hoping for one last glimpse of his selkie wife.

The Kópakonan legend begins in a similar fashion, although it deviates at a crucial point, as we shall see.[1] In the Faroe Islands, people believed seals were humans who drowned themselves at sea. Yet they were permitted to live as a human for one night every year on the thirteenth night of the year. They returned to land, removed their seal skins, and enjoyed dancing and games. One night, a young farmer on the northern island of Kalsoy, wondered if the legends were true. He left his village of Mikladalur and hid on the beach. He watched the seals arrive, who removed their skins and laid them on the rocks for safekeeping. One seal girl, Kópakonan, left her seal skin close to the farmer's hiding place. Finding her very beautiful, he stole her skin when the seal people were preoccupied with their revelry.

As the sun began to rise, the seal people retrieved their skins and swam away as seals once more. Yet the seal girl could not find her skin. The farmer crept out from his hiding place, holding the skin, though he refused to give it back. Possessing the skin gave him sovereignty over her so she had no choice but to follow him home. She became his wife, and they had several children. The farmer hid her seal skin locked in a chest to prevent her from reclaiming it, keeping the key on his belt at all times.

One day, while out fishing, he realised he had left the key in their cottage. He rushed home to find his wife gone. She had extinguished the fire and hidden the knives so their children would be safe until his return. Kópakonan had regained her seal form and swum out to sea. She was reunited with her seal partner, who had waited for her during her imprisonment on land, although

---

1 The story is related on the Visit Faroe Islands website (2016).

some believed she watched over her children from the water when they ventured onto the beach.

A few years passed, and the men of the village decided to hunt seals. The farmer dreamed of Kópakonan, and in that dream she described a bull seal and two pups, asking him not to kill them for they were her husband and sons. The farmer ignored the plea, and the villagers killed as many seals as they could. The farmer even won the bull seal and the flippers of the pups as part of his catch. That evening, a great troll stormed into the smoke room of his farm. It was Kópakonan, having taken a new form in her fury. She recognised the remains of her partner and sons and issued a terrifying curse. The men of the village would die, either at sea or by falling from the cliffs, until the dead numbered enough to link hands around Kalsoy. She disappeared when she had spoken these words. As the legend goes, men from the village still drown at sea or fall from the cliffs and those who know the story mutter that there have not been enough victims to link hands around the island. There is now a statue of Kópakonan on Kalsoy, in the village of Mikladalur, to commemorate the story.

Figure 18. The statue of Kópakonan on Kalsoy by Kallerna [CC BY-SA 4.0]. Image from Wikimedia Commons.

The legend of Kópakonan touches on the central narrative at the heart of many female selkie stories. Human men exert their dominion over the natural world by stealing the seal skins, forcing the selkies to become their wives. It was only by stealing their skins back that the selkie wives could regain their true form and disappear into the waves. There is an element of an abandonment narrative within the Kópakonan legend, where her commitment to her human children is called into question. She chooses a return to the sea, rather than the adoption of a traditional maternal role, while expecting the husband that she abandoned to spare her seal family. It would appear that her rejection of domesticity and motherhood is considered as monstrous as her shape-shifting nature. Ultimately, it is the farmer that forces both of her transformations, by keeping her in her human form and killing her selkie family. The only heinous behaviour we see occurs on the part of the farmer, yet it is Kópakonan who is physically depicted as being monstrous.

Barbara Creed first mapped the concept of the monstrous feminine, defining a monster "in terms of her sexuality" (Creed 1993: 3). Here, Kópakonan only becomes monstrous once she rejects her human form. This movement from "wife" to "monster" is made explicit through her final, and somewhat inexplicable, transformation into a troll when she issues her curse. Yet this is not the monstrosity for which she is punished, since this occurs *after* the slaughter of the seals. Instead, her husband punishes her for leaving him and returning to her original partner. Here, her choice of seal sexuality over human sexuality, enacted through the reclamation of her own skin, renders her monstrous in her husband's eyes.

Even in the legends of selkies in which the human husband has *not* forced marriage by stealing the skin, this abandonment narrative again becomes clear. These selkie wives choose to marry and yet also choose to leave, with their husbands oblivious to the true nature of their wives. Their dereliction of domestic duties sees their husbands cast as innocent victims, and it is possible these narratives provided acceptable excuses should a man's wife choose to leave the marriage. Rebranding an absent wife as a selkie who has returned to the sea invites less judgement than the potential for mistreatment on the part of the husband (McEntire 2010: 135). These departing selkies become aligned with the monstrous feminine through their refusal to adhere to gender stereotypes.

This is not always the case within selkie legends. Many tales stress what excellent wives female selkies make. Another story from the Faroe Islands sees a human man caught in a vicious storm at sea. His selkie wife dons her seal skin and saves his life. Yet in doing so, she can never return to land. She gives up her human existence with him in order to save his life. Some legends see the selkies take their children with them when they return to the sea. Other selkies leave their children on land, but continue to play with them on the beach, remaining an influence within their half-selkie offspring's life. These legends attempt to merge the supernatural with the mundane, and these selkie wives never reach the levels of power ascribed to Kópakonan. Their continued presence in the world of humans allows them a degree of flexibility around the border between this world and the supernatural realm, and thereby allows them to avoid the "monstrous feminine" label bestowed upon Kópakonan. Allan Asbjørn Jøn (1998: 97–8) even suggests that Hans Christian Andersen might have encountered the selkie story while travelling in Scotland.

For all its mythical allure, the selkie is a shapeshifter, rather than a liminal creature like the mermaid, who is both human and piscine at once. By comparison, the selkie operates in a more "binary" function, being either a seal at sea or a human on land. Clearly, there is still an element of transformation since the selkie must either remove or don their seal skin in order to change form. For Stuart C. Aitkin (2010: 14), the selkie is actually human beneath its seal skin. Here, discarding the skin allows the selkie to *choose* which form he or she will adopt. It is this conscious choice on the part of the female, combined with the ability to transform, that hints at the potential for monstrosity.

These narratives speak to the peril faced by those who are dependent upon the land and the sea for their living. The capture of a selkie demonstrates humanity's attempt at dominion over the natural world. Stealing the skin and forcing the selkie into marriage becomes a perverse means of "civilising" the wilderness, while the birth of children between the interspecies couple becomes an act of sympathetic magic. The fertility of the selkie wife seemingly ensures the fertility of the natural resources upon which isolated communities depend. The legends align the selkies with the supernatural and by extension, with the awesome power of nature and the wilderness. Allan Asbjørn Jøn repeats the Faroe Island belief that selkies are the embodiment of "the souls of drowned people" (1998: 96–7), thus linking nature and death in one figure. Some Shetland stories see selkies appear at Midsummer to lure humans into the sea. These humans never return to the land, but it is unclear as to whether they also drown. The implication is that drowning humans is the only way to form new selkies, although as the Kópakonan legend reveals, this was perhaps not a widely held belief, with the selkies able to form their own family units.

The human-seal transformation may even have a plausible explanation. Folklorist David MacRitchie (Silver 2005: 97) thought Scottish settlers mistook Finnish or Sami women for selkies due to their sealskin clothing. From a distance, it might look like a seal removing their skin to appear human. Elsewhere, some believe the sight of a distant kayaker may have looked like a seal. If you saw them haul up onto a rock and take off their sealskin cloak, a witness may become confused about their true nature. Furthermore, there is also a suggestion that the selkie story derives from an era when health conditions were less understood. A clan in the Orkney Islands had skin growths between their fingers and the resemblance of their hands to flippers strengthened their claim to be descended from selkies (McEntire 2010: 128). Much like the changeling myth and fairies, people may have "blamed" inter-marriage with selkies for atypical physical features or behaviour.

While they are not the focus of this discussion, it is worthwhile to briefly touch upon male selkies. They appear less often within legends, and they are presented as more proactive than their female counterparts. In human form, male selkies appear as handsome men and were believed to carry off married human women. Male selkies apparently preferred dissatisfied wives, which begs the question of how often male selkies became an excuse for women to abandon their husbands. One custom suggests that a woman need only shed seven tears into the sea to summon a male selkie. It is interesting to note that a woman did not need to steal his skin to enjoy an encounter.

This sharp contrast between the male and female selkie helps to underline the punishment that lies at the heart of the legend of Kópakonan. While the male selkie has amorous intentions towards women, few stories see the women come to harm. Rather, they actively seek out his company. Instead, the female selkie is often the victim of male abuses of power, a docile creature that misses her home. It is this longing for her true home that sees Kópakonan disobey her husband and choose for herself where she would prefer to live. For this simple act of personal choice, she is considered a monster and her selkie family pay the ultimate price.

Daisy Butcher

The Little Mermaid (Hans Christian Andersen, 1837)

Feminine Magics

The Little Mermaid (1837) is perhaps Hans Christian Andersen's most well-known and enduring tale. It tells the tragic story of a young mermaid who falls in love with a human prince and offers her voice to the sea-witch to transform into a human. The character of the sea-witch is generally understood to be as libidinous as she is repulsive. She serves as an early nineteenth-century manifestation of a much older misogynistic archetype of the witch character, bound up with hagsploitation and used to depict female sexuality as something inherently disgusting, diseased and abject. In this article, I will explore the ways in which tentacular symbolism is used to position the sea-witch as an embodiment of transgressive female sexual pleasure and monstrous female agency and how she is deserving of redemption.

The Little Mermaid turns to the witch for help after being dissatisfied with the advice of her grandmother, known as "the old dowager" or "aged mother of the Sea King" (Andersen 1837: 17, 2). Filling in for the Little Mermaid's dead mother, the grandmother is a positive, caring and obedient older female figure in direct contrast to that of the sea-witch later in the story. Turning away from the care and guidance of her grandmother, the Little Mermaid symbolically moves away from conservative womanhood and sexual restraint towards transgressive femininity and sexual pleasure by choosing the sea-witch as a mentor (see Cashdan 1999). Andersen writes:

> And then the little mermaid went out from her garden, and took the road to the foaming whirlpools, behind which the sorceress lived. She had never been that way before: neither flowers nor grass grew there; nothing but bare, gray, sandy ground stretched out to

> the whirlpool, where the water, like foaming millwheels, whirled round everything that
> it seized, and cast it into the fathomless deep. (Andersen 1837: 17)

The whirlpool is of particular interest here as it conjures imagery of deadly maelstroms dragging ships down into the depths of the sea as a symbolic descent into the underworld. In nautical folklore, there is often a kraken at the heart of the maelstrom or perhaps the monstrous Scylla from Greek mythology awaits, like a spider at the centre of its web, with dog-heads sprouting between her legs. With Kraken, Scylla and Melusine imagery combined, the whirlpool can be read as a destructive maw and abyssal iteration of the *vagina dentata*, a femininsed site of sexual danger.

Through the terrified eyes of the Little Mermaid, it appears that the area surrounding the sea-witch's house is not as devoid of life as it first appeared:

> [A]ll the trees and flowers were polypi, half animals and half plants; they looked like
> serpents with a hundred heads growing out of the ground. The branches were long slimy
> arms, with fingers like flexible worms, moving limb after limb from the root to the top.
> All that could be reached in the sea they seized upon, and held fast, so that it never es-
> caped from their clutches. (39–40)

These wormlike "polypi" are described as slimy limbs and as such can be read as tentacles (see Figure 19). These tentacles serve as extensions of the sea-witch herself and are depicted as uncontainable animal-vegetable hybrids, thus cementing the understanding of the sea-witch as a radical and nonconformist figure who transgresses boundaries.

This transgression of boundaries is something that can be seen particularly within tentacular eco-horror as nature snakes its way out of its confines. Dawn Keetley breaks down tentacular eco-horror into two main movements, the first being "an impersonal nature that asserts its own life with rustling insistence," and the second "is the human's entanglement with this nature. Nature reaches out its tentacles, rendered all the more alien in their startling agency" (Keetley 2021: 30, 31). Applying these notions to the sea-witch, her use of tentacles allows her to forcibly grasp agency in a male-dominated environment and blur the borders of gender, humankind and nature, the self and the Other.

Another layer of meaning concerning tentacles and women comes from the work of Eric Neumann, a Jungian, psychoanalyst who wrote extensively on fear of the feminine. He writes:

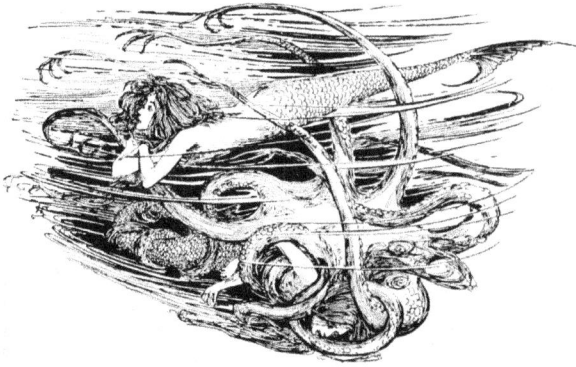

Figure 19. One of Helen Stratton's Illustrations which first appeared in *The Fairy Tales of Hans Christian Andersen* (1899), published by J. B. Lippincott in the public domain.

> The Great Mother in her function of fixation and not releasing what aspires toward independence and freedom is dangerous ... To this context belongs a symbol that plays an important role in myth and fairy tale, namely, captivity. This term indicates that the individual who is no longer in the original and natural situation of childlike containment experiences the attitude of the Feminine as restricting and hostile. Moreover, the function of ensnaring implies an aggressive tendency, which, like the symbolism of captivity, belongs to the witch character of the negative mother. Net and noose, spider, and the octopus with its ensnaring arms are here the appropriate symbols. (1963: 66)

Not only, then, do tentacles equip the sea-witch with monstrous agency but she also incites fears of entanglement with her ensnaring polypi. This could be both as a dangerous *femme fatale* or siren who lures men astray with her brazen sexuality or, equally, in tempting good girls over to her sexually deviant ways. Neumann juxtaposes kraken and whirlpool imagery with the smothering, vice-like grip of the Terrible Mother. Moreover, reading the tentacle as a metaphor for social reach/mobility and the sea-witch's identity as a (and affinity with) sea creature[s] reinforces her transgressive credentials due to the symbolism of the Ocean as a fluid space, lacking a fixed state.

After passing through the tentacled forest, the Little Mermaid meets the dreaded sorceress as Andersen describes in the following monstrous tableau: "There sat the sea witch, allowing a toad to eat from her mouth, just as people sometimes fed a canary with a piece of sugar. She called the ugly

Figure 20. One of Helen Stratton's Illustrations which first appeared in *The Fairy Tales of Hans Christian Andersen* (1899), published by J. B. Lippincott in the public domain.

water-snakes her little chickens and allowed them to crawl all over her bosom" (Andersen 1837: 42). It is clear that the sea-witch shows kindness to tradition-ally "ugly" creatures such as toads and snakes which are, unfairly, demonised as vermin. These two animals in particular, toads and snakes as biblical symbols of pestilence and phallic imagery, respectively, and have a long and deep-rooted association with witches, supposedly to augment their hideous and repulsive femininity. The toad eating from her mouth and snakes crawling over her breasts have striking erotic connotations as a grotesque and decadent display of cunnilingus with an emphasis on female sexual pleasure as disgusting, hor-rifying and unnatural (see Figure 20).

Reading the polypi in her forest and the sea snakes on her breasts as ten-tacular, it is interesting to appreciate the sexually charged octopus iconog-raphy present within Andersen's sea-witch. In Western imagination, Richard Schweid argues that the octopus typifies "Eros and Thanatos, desire and death," strengthening its connection to that of *femme fatale* witch figures (2014: 126).

This juxtaposition of female sexual pleasure and writhing tentacles in Andersen's *The Little Mermaid* bears striking resemblance to that of Katsushika

Hokusai's *The Dream of the Fisherman's Wife* (1814) (see Figure 21). The image depicts a woman with her legs open as one octopus performs oral sex on her and another caresses her breasts. While it is generally understood as a celebration of female pleasure, sexual fantasy and enjoyment, many male Western observers during the nineteenth century (such as Joris-Karl Huysman) interpreted it as a rape scene, the woman's facial expression showcasing pain rather than heights of sexual pleasure (Schweid 2014: 128). With this in mind, the sea-witch scene in *The Little Mermaid* can be understood from the context of a Western male gaze which both scandalised and repressed female sexuality. As a result, the Western view positioned sex as something done to/inflicted onto women rather than something women enjoyed or actively participated in and, furthermore, that any depiction of female orgasm/pleasure is positioned as repulsive and unnatural as exemplified by Andersen's sea-witch in the aforementioned scene.

However, it is possible, and I will also argue, that Andersen's description of the sea-witch can be read in the same way as Hokusai's *The Dream of the Fisherman's Wife,* and that the Danish storyteller inadvertently recreated a scene which celebrates female orgasm. The sea-witch can be reclaimed as a nonconformist figure who flagrantly disregards Western modesty culture, and openly exhibits the pleasures of the female body despite patriarchal attempts to position it as something shameful or disgusting.

This reclamation of the sea-witch as a sexually transgressive figure also feeds into psychoanalytic symbolism surrounding the octopus. For example, Jacques Schnier, a psychoanalyst during the mid-twentieth century, argued that the negative perceptions of the octopus in Western thought were because it was "viewed by men as a symbol of a woman with a penis" (paraphrased by Schweid 2014: 133). The sea-witch, bound up in all the life-giving, death-dealing and unconfined nature of ocean imagery, therefore, is a manifestation of gender fluidity which is further reinforced by her affinity for hybridised forms and tentacles. This transgression of boundaries situates the sea-witch as a potential figure of female defiance as she refuses to be contained in traditional gender roles of wife/mother and instead, uses her tentacular reach to grasp and influence the male-dominated world by symbolically appropriating the phallus.

Moving on to the Faustian-style pact between the Little Mermaid and the sea-witch, I further argue that the sea-witch is primarily advising the mermaid

Figure 21. *The Dream of the Fisherman's Wife* by Katsushika Hokusai (1814). In the public domain.

of the consequences and price of her decision to become human. She makes no effort to conceal the true nature of the transformation or the cost the Little Mermaid must pay by clearly stating "at every step you take it will feel as if you were treading upon sharp knives, and that the blood must flow. If you will bear this, I will help you" (Andersen 1837: 44). There is no trickery here, the sea-witch is not seeking to dupe the Little Mermaid out of her voice and seeks for informed consent before she proceeds to create the potion. The emphasis on pain is important as it is in direct regard to a lower body metamorphosis which serves as a metaphor for young girls' experience of menarche and puberty more broadly, or indeed the pain experienced during a first sexual encounter.

As well as explaining the painful side effects, the sea-witch goes on to explain the permanent repercussions of such a decision. She advises the Little Mermaid to "think again" as if warning her against being overly impulsive, and continues:

[F]or when once your shape has become like a human being, you can no more be a mermaid. You will never return through the water to your sisters, or to your father's palace again; and if you do not win the love of the prince, so that he is willing to forget his father and mother for your sake, and to love you with his whole soul, and allow the priest to join your hands that you may be man and wife, then you will never have an immortal soul. The first morning after he married another your heart will break, and you will become foam on the crest of the waves. (Andersen 1837: 45)

It is obvious then, that the sea-witch clearly defines the risks involved in the Little Mermaid's wish, including the fact that she must sacrifice her voice in payment for the spell. She equips the Little Mermaid with the information necessary to make her own informed decision and, crucially, is the only character in the text to allow the Little Mermaid the power to choose and to control her own fate and body. The sea-witch serves, therefore, serves as a "wise woman" character and someone for the Little Mermaid to speak to about her newfound womanly urges and desires. Wise women, of course, were medical women, often midwives, who were able to advise on matters of fertility who were often unfairly demonised as "witches" in league with the devil in Early Modern Europe. In this respect the sea-witch is no more evil than a doctor who explains the options and their side effects to a patient and who then carries out whatever informed decision the patient makes.

After cleaning out her cauldron using sea snakes as a scourer, the sea-witch sacrifices her own blood in order to create the Little Mermaid's potion. Andersen writes how she "[pricked] herself in the breast, and let the black blood drop into it" (Andersen 1837: 47). Once the Little Mermaid possesses this potion mingled with the witch's blood, she is able to pass freely through the forest of polypi without being attacked by the tentacles. As a result, the Little Mermaid's quest for the potion serves as a metaphor for defloration as passing through the phallic tentacled space coupled with the presence of blood within the potion carries undertones of hymeneal breakage due to penetration. This is strengthened by the fact that the polypi/tentacles cower from the potion and are no longer interested in dragging the Little Mermaid down as she has lost her previous innocence and its associated value.

The cephalopod imagery surrounding Andersen's sea-witch was clearly not lost on The Walt Disney Company with their characterisation of Ursula in *The Little Mermaid* (1989). Rather than the ambivalent figure in Andersen's tale,

Disney recasts the sea-witch as the film's main antagonist. Famously modelled after the drag queen Divine, the film continues the strong associations present in Andersen's tale between the sea-witches, octopi and gender nonconformity. On Ursula's avant-garde sexuality, Marina Warner writes that Disney's "Sea Witch ... expresses the shadow side of desiring, rampant lust; an undulating, obese octopus, with a raddled bar-queen face out of Toulouse-Lautrec and torso and tentacles sheathed in black velvet, she is a cartoon Queen of the Night, avid and unrestrained, what the English poet Ted Hughes might call 'a uterus on the loose'" (Warner 1994: 403). Warner highlights Ursula's character design as a manifestation of female lust even so far as to compare her to a personified uterus, with all the *vagina dentata* implications as a result of her kraken-like form. Like Andersen's sea-witch, Ursula forces the reader/viewer to confront female sexual pleasure and bodily autonomy much in the same way as Hokusai's fisherman's wife.

To conclude, while the Little Mermaid is eager to assimilate into society and to mutilate/castrate herself to be a dutiful wife to the prince, the sea-witch is a strong, defiant and knowledgeable woman who refuses to conform. Through embracing nature and her affinity with all things tentacular, the sea-witch is both a male nightmare of monstrous female agency and a female fantasy of empowerment, pleasure and freedom. Tentacular horror has become a staple of eco-horror and the tentacle is well suited to ecofeminist readings exploiting both ecophobic and gynophobic fears. As hinted earlier, the similarities between Hokusai's *The Dream of the Fisherman's Wife* and Andersen's sea-witch could suggest a knowledge of the former in the latter. However, Andersen's response to shared themes is oppositional to the Japanese artist's, overtly demonising female sexual fulfilment instead of celebrating it. Ultimately, I have sought to reclaim this scene as an act of ecstatic sexual defiance. The sea-witch's defiant nakedness and enjoyment of sex confronts the male gaze and challenges its fetishisation of female beauty and youth in the process. This further sees the sea-witch as a Baba Jaga-like figure who does nothing without full, informed consent and in doing so, grants the Little Mermaid her own bodily autonomy. Therefore, she requires a feminist reclamation in the same way that the monstrous women such as Lamia and Medusa have been redressed and even celebrated as empowering symbols of female sexuality and nonconformism.

Martine Mussies

#Posidaeja (Efa, 2021)

Until recently, it seemed that many mermaids were doomed to lead a tragic life, stuck between longing and belonging, in a space that Homi Bhabha calls "Unhomeliness" (1994: 9). European mermaids seemed helpless at the mercy of the vagaries of life, tossed about by the choices of male figures in their respective narratives. Authors and artists alike framed the water woman in different versions (including Andersen's) as a tragic figure, commonly a victim of love, often relegated to the sea. This image of the mermaid is apparently so strongly rooted in our common consciousness, that it is even used by feminist scholars like Dorothy Dinnerstein (1976). As I hope to demonstrate in this essay, the mermaid has not always been a victim of her fate, and her depiction is subject to change (Mussies 2022). With the turn of the twenty-first century, various mighty mermaids and related water women are making their comebacks online, on the "wonderfully sprawling repository of arcane fictions and crypto-everything" as Sheila Hallerton (2016) describes it.

The internet is an endless ocean of creativity, and new media have fostered a "phenomenally increasing proliferation of fairy-tale transformations in today's 'old' and 'new' media," (Schwabe, 2016: 81) and as further described by Halerton:

> Its fragmentary and often inter-generative texts thrive and gain momentum with the slightest (and often most erroneous) of pretexts, generating threads of online mythology that variously intersect with older folkloric and mythological stories or else develop independently. (2016)

Reclaiming the Feminine

As for the recent water women, many of their identity markers can be traced back to Greek antiquity, as the precursors of present-day mermaids might be surprisingly powerful. Often, contemporary depictions of mythological women are hybrids of earlier types, adapted to their new context, but with recognisable traces of their origin, made explicit in the hashtags that accompany the modern fan art as breadcrumbs: #Atargatis, #Nehallenia, #Achtamar and so on. This also counts for #Posidaeja, the 2021 case study of this essay.

Posidaeja (2021)[1]

By surfing the nooks and corners of the internet looking for fanart about Posidaeja, one comes across many similar drawings. Of course, one artist is a bit further in his or her (mostly her) development and also the techniques and working methods differ (from pencil drawings to digital animation films). The identity markers of this sea goddess are however quite consistent and can thus also be traced in the digital artwork #Posidaeja. This work was created by a Russian fan artist under the pseudonym of Efa and posted and reblogged on many social media websites, such as Tumblr, DeviantArt and WeHeartIt.[2] In addition to the hashtag containing the title of the work, Efa and her rebloggers have also added all kinds of other hashtags and explanations.[3] Sometimes this is about the artwork itself – like #blue and #digitalpainting – but sometimes also about the mythology behind it – like #seagoddess. The latter category also includes hashtags with names of similar sea goddesses such as Posidaeja. The descriptions include Amphitrite and Salacia.

1   The research presented in this chapter was partly based on my 2022 article "Posidaeja and Mami Wata: The online afterlives of two mermaid goddesses" in *Shima*. DOI 10.21463/shima.175.

2   On Instagram, Efa is known as @efa_finearts and @efa.mari.

3   An image can be found at this authors website: http://martinemussies.nl/web/another-day-another-offering/.

In the art, a sea goddess is shown with her queenly features such as wearing a crown, which symbolises the authority she has over the sea. She is also holding a trident that, according to Greek mythology, was the symbol of ruler of the sea, in addition to its power that gave the owner control over the sea. (In many other cases she is riding a chariot as well.) The eyes are shown as reflecting the sea and its sinuous waters to signify the area they rule, the sea itself, and how she is in control of it. The sea goddess is showcased bearing over a turbulent sea with her face calm, allowing the chaos around her to show the superiority and power she has over the waters. The queenly robes represent her position as royal – apparently alone as there is no depicted god of the sea – and the respect she is due as a leader. The robe further enhances her beauty, as a more-than-human entity, a very vital characteristic of a mythical goddess. The choice for blue as the dominant colour – in her eyes, hair and tonal palette – could be seen as a thematic colour, to enhance the actual (reflective) colour of the sea and waters.

Although at first glance, this image rather emphasises mainstream female ideals (such as whiteness, thinness, long hair, etc.) the fannish texts that accompany it carry emancipatory potential. With this work of fan art and the accompanying hashtags, Efa has answered the call of professor Barbara Olsen (2014), who called out for help "Cherchez la femme." Her appeal to search out unknown women at the heart of a mystery, has much relevance for the ongoing research of Greek antiquity, where, in her words, "so often women have been shrouded in myth, notoriety or obscurity." I hope to make a contribution to her work by examining these special water women – nowadays called mermaids – from Hellenic and Hellenistic civilisations.

The Historical Posidaeja

Posidaeja is one of the earliest traceable mermaid-like figures, a goddess who was important in the Mycenaean state of Pylos (1600–1100 BC). Nestor's kingdom of "sandy Pylos" was an important centre, as recalled by Homer in the Odyssey (XVII 108–12). The Pylian deity appears to be a female version of Poseidon. The theology of Poseidon and Neptune may only be reconstructed to some degree, since from very early times they were identified as one and the same god.[4] Such an identification may well be grounded in the strict relationship between the Latin and Greek theologies of the two deities (see Showerman 1901: 223). The name of the sea-god Nethuns in Etruscan was adopted in Latin for Neptune in Roman mythology, while linear B inscriptions on clay tablets show that Poseidon was already venerated at Pylos and Thebes in the pre-Olympian Bronze Age. The origins of Poseidon's cult can be traced back to the Mycenaean period (Mylonopoulos 2012), and he was later on integrated into the Olympian gods as being a brother of Zeus and Hades. In the Greek-influenced tradition, Neptune was the brother of Jupiter and Pluto; the brothers presided over the realms of Heaven. Again: a very masculine mythology. But Richard Vallance, a specialist in Mycenaean Linear B, states that when thinking about the Gods worshipped at Pylos, "it is of great significance that there are both masculine and feminine versions of each god, and [that] even where there is no masculine, there is always a feminine, because the Minoan and Mycenaean religious hierarchy is matriarchal" (2016). This can be traced in the surviving clay tablets from Pylos (Bennett 1995).

In the Minoan and Mycenaean civilisations, writing has not been observed for any use other than accounting and where clay tablets served as inventory lists, labels and summaries.[5] In our quest for the mermaid, an interesting example is the Pylian tablet known as PY Tn 316. It is written in Linear B – the oldest language we can read – and records golden "Mycenaean goblets" and "Minoan chalices," dedicated to various deities (see Cook and Palaima 2016). As Joann Gulizio explains, "Tn 316 contains a strikingly high number of female divinities" (2000). It mentions, among others, po-ti-ni-ja, di-u-ja, e-ra and: po-si-da-e-ja, female Poseidon. This possible precurser of Amphitrite appears in line four, which reads: ma-na-sa, AUR *213VAS 1 MUL 1 po-si-da-e-ja AUR

---

4   Poseidon's presence in the lectisternium of 399 BC is a testimony to the fact.
5   In this cultural region the tablets were never fired deliberately, as the clay was recycled on an annual basis. However, some of the tablets were "fired" as a result of uncontrolled fires in the buildings where they were stored. The rest are still tablets of unfired clay, and extremely fragile; some modern scholars are investigating the possibility of firing them now, as an aid to preservation.

*213VAS 1 MUL 1. According to Olsen, this could be translated as "Given to ma-na-sa with Gold vessel; given to po-si-da-e-ja along with Gold vessel" (Olsen 2014: 285).

But what about the people that are depicted? Various scholars believed that the men and women drawn near the golden vessels were human sacrifices (see Buck 1989: 131–7; Hughes 1991: 199–201). But recently, scholarship has explained the people in the tablets as being priests and priestesses who perform the ritual of offering the gold (see Sacconi 1987: 552–4). Richard Vallace is not so sure at all that the Minoans ever sacrificed humans, as they "were much too civilized and peaceful to indulge in such barbaric acts" (from personal correspondence with the author). As explained, the so-called "evidence" – painted people near the offer gifts – is very slim. Perhaps it is possible that the Mycenaeans may have done this, as they were quite warlike, but both societies were essentially matriarchal, which often precludes human sacrifices.

Mermaids' Epithets

As mentioned above, the 2021 piece of fan art #Posidaeja was also hashtagged #Amphitrite and #Salacia. In line with the statements of Sheila Hallerton (2016) that we started this exploration with, this fannish depiction also engages in re-mix culture, by blending similar elements from different historical contexts into one digital image. The online fannish depiction thus mirrors the Greek epic, where the gods often had a fixed epithet, a byname, or a descriptive term (word or phrase), accompanying or occurring in place of a name and having entered common usage. The epithet can be described as a glorified nickname, for example, when talking about the fleet-footed Achilles. Many gods and goddesses have a number of epitheta. Such as Athena with the sea green eyes, protecting the city as Polias, overseeing handicrafts as Ergane, joining battle as Promachos and granting victory as Nike and so on. A deity's epithets generally reflect a particular aspect of that god's essence and role, for which his/her influence may be obtained for a specific occasion.

In ancient Greek mythology, Amphitrite was the wife of the god of the sea, Poseidon. Being his wife, she was referred to as the goddess of the sea. She was the daughter to Nereus and Doris and attributed to the Olympian pantheon due to her union with Poseidon. She had authority over the sea and its creatures and was even thought to give birth to sea monsters such as the Kraken. In contrast to Poseidon, who was considered violent in his ruling, Amphitrite was calm and gentle. Her Homeric epithet was Halosydne (Greek: Ἀλοσύδνη), meaning "sea-nourished" (Vollmer 1874). Salacia was her counterpart in Roman mythology, a primordial sea goddess, the goddess of saltwater. Since Salacia was the wife of Neptune, her name was generally associated with salt and the sea (Demicheli 2007). Together with Neptune (the Roman version of Poseidon), Salacia presided over the depths of the ocean.

Next to #Amphitrite and #Salacia, there are many more epitheta for the earliest mermaid-like figures that can be encountered in present-day fannish depictions. Derketo, the ancient Greek version of the Assyrian Goddess Atargatis, is associated with powerful symbols of fertility and sensuality, with fish-friends and with jewels/pearls. She is also connected to Artemis, who was (among other roles) a goddess of the moon, a goddess of maiden dances and songs and a goddess of fish and fishing. The moon and maiden dances are also exponents of fertility rites and symbols of feminine strength and power. In the person of Artemis, a link is thus made between the fish figure and the female. The mermaid is not only powerful as a goddess (because she is a goddess), but she also represents the power of women (the female).

Sometimes there are features on the drawings that are "adult content." It would be too easy to dismiss the (severed?) penises on these works of art as adolescent and insipid. Because also these elements have their precursors in ancient history. The loose male genitalia resonate with the stories of Atargatis, a mermaid-like goddess documented from the fourth century BC in northern Syria. Modern scholars have known her since 1933, when archaeologist Michael Rostovtzeff described the "last German expedition to Palmyra" in 1917. In his 1933 article, he writes about the discovery by Professor Wiegand of the foundations and of some architectural fragments of a large and beautiful temple to the south of the main street of Palmyra. In the South-West corner of the temple, Wiegand recognised the end of the body of the fish, leading him to identify Atargatis, the goddess we now know that was worshipped there as she was also depicted on the coins of Palmyra (Wiegand 1932). Atargatis was a fertility, and mother goddess, who transformed herself into a mermaid out of shame for accidentally killing her human lover. Her original Syrian name Atar-ata is explained in the literature as a combination of the components Astarte and Anat.

In the time of the sovereignty of the Seleucidae (312–63 BC), the worship of Atargatis had spread among the Greeks, and from them found its way to Rome. During the Hellenistic era (323–31 BC), the Romans worshipped her as *Dea Syria*. Geographically, we can connect Atargatis to at least three temples, in Hiërapolis, Palmyra and Qarnaim. But she also had a temple in Rome during the days of the Empire, and in other parts of Italy, and still farther west. Hiërapolis, the former Bambyce, in northern Syria was the main centre of her cult. Here she shared a huge temple with her consort Hadad, one of the earliest examples of a merman. In *De Dea Syria* (second century AD), the author Lucien of Samosata reported that he saw a statue of a Phoenician goddess who was a mermaid and was called Derketo (Strong 2015). The credibility of this treatise has been debated, but according to Lucinda Dirven (1997), it is likely that Lucian is not the author, so this claim may in fact be more accurate than was previously supposed. According to *De Dea Syria*, the worship of Atargatis was of a phallic character, with orgies, with votaries who offered explicitly male figures (made of wood or bronze) and with very large phalli before the temple, which were climbed and decorated in ceremonies. All to worship and celebrate the fertility brought by this powerful mermaid goddess. Near the temple was a sacred lake, where her fish-friends lived, who were ornamented with jewels and/or pearls.

Conclusion

According to Haasse, after the tipping point of the Middle Ages and the Modern Era, mermaid myths were debunked, and mermaids lost their powers and agency. In later traditions, "the mermaid shifted from a powerful, siren-voiced temptress to a mute, shadowless, soulless creature in the need of human [male] aid" (Jarvis 2008: 621). But just like her sisters, the other Hellen(ist)ic precursors of modern mermaids, Posidaeja remained forgotten or passed over by the currents of popular culture. However, many of her and her sister's identity markers resurfaced and found their way to twenty-first-century online fan art, such as the 2021 digital art work #Posidaeja. As Tamra Andrews explains:

> The mermaids patterned after Atargatis's prototype had numerous attributes that identified them with the sea. They had long flowing hair, often green like seaweed or like rays of sunlight falling on the water, and they often held mirrors to reflect the light of the moon and to identify them with the moon's control over the tides. Mermaids lived in undersea homes made of pearl and coral but were often seen basking on the sun-drenched rocks, gazing into their mirrors and combing their long, wavy locks. (2000: 118)

But unlike the majority of their successors, the precursors of modern mermaids turned out to be surprisingly powerful. In 1600–1100 BC, Posidaeja was a goddess, worshipped and adored, as people built her temples and dedicated offerings to them. In 2000 CE, she came back in online fan art, just as beautiful and powerful as three and a half millennia before. We searched for the woman and found mighty mermaids.

Phil Fitzsimmons

Underwater (William Eubank, 2020)

Introduction

This chapter focuses on the 2020 movie *Underwater* as an exemplar of how Homer's "Song of the Sirens" has emerged as a commentary on the current global silence regarding the current environmental crisis. Found in both the *Iliad* (c. 800 BC) and *Odyssey* (c. 800 BC) they unpack the aftermath of the Trojan wars in which Odysseus recounts his decade-long return to his island home of Ithica. Prior to departure, the sorceress Circe warns Odysseus he will encounter the Sirens, mer-women, who lure men to their deaths in the ocean through their voices. However, as Micheal Bull points out the Siren narrative has been misconstrued as it is about misogyny that "silences the voices of woman" (2020: 10).

Underwater cinematically carries forward Bull's assertion, fleshing out the impact of the masculine drive "to possess, and dominate the physical spaces of the past and present" (Baaqueel 2019: 83) and its impact on the climate crisis. In effect, this movie is an eschatological commentary on industry's silence, and disconnectedness to the destruction of the environment.

Sirens in the Schema of Things

Homer portrays the Sirens as monstrous nymphs perched on an island, "lolling in their meadow, round them heaps of corpses rotting away, in rags of skin shriveling on their bones" (Homer 2006: 48–57). In myth an island is marked as male, enabling "the propagation and maintaining of a perfect masculinity" (Loxley 1990: 117). However, a meadow is coded as female in this narrative perspective. Hence, the site can be seen to represent the feminine as being in harmony with the earth and sea environments, while the corpses around them imply the failed patriarchal attempt at subjugating the environment through "mastery and domination" (Russell 2017: 207).

Research indicates that the Sirenic narrative arose within the Ishtar myth around 3,500 BC, and for millennia afterwards was synonymous with this goddess. As a double-coded figure, Ishtar represents both love and war emerging in a period of existential uncertainty when changes in the natural world appeared to correlate with challenges to patriarchal dominance. According to Irvine, the existential dilemma of that epoch was rapid climate change, "with rising temperatures melting the retreating ice sheets of the northern hemisphere causing global flooding with a growing population unable to spread out" (Irvine 2020: 29). Along with the connection between birth and amniotic fluid it would appear that this inundation event gave rise to the ocean "representing the primordial intrauterine world of maternal darkness, water, and death" (Callaghan 2013: 157). This connectivity still narratively positions the female as such, with the patriarchy superficially connecting to the natural world or "pretending to be linked to the natural setting" (Fitzsimmons 2016: 109).

Siren Imagery Resurfaces

Underwater is located entirely in a mining zone at the bottom of the Mariana Trench. Operated by Tian industries, the company aims to drill seven miles through the sea floor searching for minerals. As a result, unknown creatures have been released, attacking the site leading to its collapse. The central protagonist is a mechanical engineer, Norah Price (Kristen Stewart), who as a woman is also marked as "a sign of monstrous difference" (Braidotti 2011: 226).

The Feminine in Paratextual Frames

The movie commences with three paratextual signatures of the monstrous feminine. It is important to note that from this point there is an absence of time and natural light. Within monster narratives this suggests a general absence of relationships, and a coming confrontation with the horror of self that leads to a descent into death. Paralleling the indeterminate modality of Homer's texts, it is unclear who or what will die.

In the first paratextual section, the title of the film is set in the bottom left hand corner of the screen below the horizon line. In this position in a black background the semiotic implication is that narrative is enmeshed in "the primordial feminine abyss" (Callaghan 2013: 211). As a spatial signifier this is a double marker, synchronously representing both "the east gate of birth and closer of wombs" (Neumann 2015: 170).

In the ensuing section, the mining company is named as Tian. As a reference to "the oldest symbol of heaven in Chinese mythology" (Roberts 2009: 23), the question of cultural ownership comes to the fore. In this conjoined opposite, or Ying-Yang metaphor, hell is a component considered to be purgatory in the depths of the earth. This figurative space is also a place of decision in which dire choices have the possibility of being redeemed. Creed likens this space to "a mythological figure of woman" that is both the Eternal Mother and "the monstrous vagina, and the origin of all life threatening to reabsorb what it once birthed" (2015: 56).

Morphing into a third reference, newspaper-like texts appear alluding to a disaster at the Tian site and sightings of unknown creatures. The entire mine is then mapped on screen as a uterine-fallopian outline. This schematic resting on the earth's deepest point is a visual narrative of masculinities deepest fears: the saline-filled womb. Aydemir contends that this biological metaphor has a broader meaning in that it is representative "of the sea of pollutants in which masculinity finds itself at risk of dissolving in is not only environmental (the world) and physical (the womb), but also cultural" (2007: 16).

This paratext implies that this mining company, and by implication patriarchy, has the power to displace heaven from its natural setting, and with impunity set it in the entrance to the "mouth of hell, the *vagina dentata*" (Creed 2015: 6). Holistically, these three markers of territory carry with them all the symbolic relationships of dominance, in tandem with the emblematic features of the patriarchal fear of womb encasement, birth and complete dissolution of all they have constructed. It could also be argued that this intertextual layering is a diegetic Sirenic call to resist the global destructive forces and listen and "trust their warning sounds" (Bull 2020: 5). In the silence of the seafloor, this is the loudest call to listen to the representation of "woman's time, of cycles, gestation and the eternal recurrence of a biological rhythm which conforms to that of nature" (Burrows 2016: 107, 111). Echoing this sense of cycles, *Underwater* moves through plot repetitions, intertextually echoing the biblical and mythical symbolism in which the number three represents the unholy trinity of the Beast, the "monstrous corporality" (Mazurek 2014: 63) which ushers in the final destruction of the earth.

Thus, the appearance of monsters, who "notoriously appear at times of crisis," (Cohen 1996: 21) is a portent of collapse. Hence, the monstrous incursion within "the devil's gateway" (Creed 2015: 6) indicates the equilibrium of the natural environmental is at the eschatological tipping point.

From Paratext to Eschatological Imagery

The first narrative section takes up the first thread of paratext through the use of a visual downward vector. Following a tether into the oceanic darkness, this is the first in a metonymic chain of references to "the hydraulics of ejaculation" (Badley 1996: 92). This plunge ends at a mining structure, echoing the ideal of an ovum. Taking the lead in the storyline Price simultaneously becomes connected to the notion of the feminine as primordial darkness and an inversion or a manipulation of female agency. This is an early indicator that possibly only she can understand the emblematic ideals of this location and become the feminine restorative agent in regard to the last gasp of the feminine "primordial energy" (McLauglan 2012: 52), or she is possibly the feminine destroyer.

Once inside the mining building a 360-degree pan reveals an empty labyrinth of corridors devoid of human presence. As (Creed 2015: 9) contends, this motif is one of the oldest "allusions to the monstrous vagina." With the drop to the bottom generating the notion of insemination, the visual focus on an empty space on the seabed continues the spatial metaphor of man being swallowed up by the same apparatus that sent him forth into the world. Price's doubleness is again recreated, saviour or destroyer, allowing "for identity to merge" (Botting 2008: 8). Again, this sets up the critical question: Will Price become one with the figurative site of reconciliation with "the Great Mother" or enter the female emblem who is also "the abyss of death, threatening to fill the world with water?" (Neumann 2015: 187).

The 360-degree sweep ends in an open doorway of a communal bathroom, where Price is seen alone cleaning her teeth and wearing only a bra and tight-fitting pants. Her short blonde hair, lack of clothing, hint of androgyny and "whiteness, youth and thinness" fits with Farrimond's (2018: 9) definition of the current cinematic sense of the femme fatale. In this gender shading her role in the company, becomes a representation of the tensions between "femininity and the male representatives of patriarchal control" (Farrimond 2018: 80). Her sense of place in this tension becomes a haunting and silent sense of loss, added to by the absence of bathroom markers delineating places of male and female elimination of the "abject such as human waste" (Kristeva 1982: 2).

In the silence of reflexivity, the ideal of loss is continued through a voice-over of Price's thoughts: "After months underwater you lose all sense of night and day. There's only awake and dreaming, and they're not easy to tell apart" (Eubank, 2020). It's also clear she was in a failed relationship before coming to the mine site, in which her male partner was an optimist, while she was the opposite. She admits to herself "there is comfort in cynicism. There is a lot less to lose" (Eubank, 2020).

In this place of developing unreality Price has already lost more than she realises. In a container where the huge excess of ocean pressure appears to match the social-emotional and gender diminution, Price is flagged as possibly experiencing seepage into internal masculisation. Although already "abstracted from man" (Derry 2009: 33) by way being female, she is also an emblem of reproduction who seems to becoming affected and infected by the masculine institution.

Once articulating there is not a male in her personal life the façade of the psychosocial walls of containment suddenly collapse. Immediately Price feels a drop of water falling from the ceiling, followed by the implosion of the walls of the corridor allowing seawater to gush in. Reflecting the porous inadequacy of the man-made boundaries between the ocean from the deep-sea intrusion of Tian, mythically the ocean appears to be reclaiming what was hers. This seems to include Price.

Signalling that the tipping point of the climate crisis has been reached, Price and an unknown male are swept into a saline melee. They barely manage to get to a control hub, returning once again to a space highly symbolic of womb-like containment. Price is aware of yet another male behind her being washed down the passageway, but on entering she is forced to close the hatch in order to save their lives. Yet again Price enters the "interuterine life, of existence in the womb" (Ebbatson 2016: 71). Realising the still present danger, she and her companion once again move on, shifting out of a space symbolic of the male urge to "contain, control and delimit" (Cremin 2021: 149).

In the briefest of synopses, Price and the other male survivor manage to crawl out once again through collapsed rubble-filled tubes. She digs another male out of the rubble and these three then find their way to an entry point of another control room that has three other survivors. Through Price's computer skills, she is able to open the doors into the midway pod and then source a schematic of the drilling site. She realises the small group of survivors have to repeat their previous escape, and so once again have to drop to the sea floor in pressurised suits and walk to another control room in order to exit to the surface via escape tubes. Trudging along the seafloor they encounter a monster eating a dead miner. One cannot help noticing the interrogative facial expression of Price, the look of recognition of the monster, the silence and the darkness as she realises that the creatures have overtaken the construction site. The primordial appears to be far stronger than the patriarchal. Entering another submerged corridor, they realise something is following them. This unknown creature catches up and kills the last male about to enter the second pod.

This first excursion into the sea depths is reminiscent of the biblical eschatological narrative of Jonah's plunge into the sea and the experience of being in the belly of the whale. This cinematic narrative of enclosure also fits within the long-held Jonah trope of the sea floor representing "an inquiring for the ruinous door of the womb" (Nordin 2006: 188). However, just as in the Jonah narrative, embedded in this metaphor are "the juxtaposing images of ruin, estrangement and hope" (Nordin 2006: 188). Price is now emblematically straddling the focus of redemption of the natural world, as well as the inverse in which "human artifacts, products of culture may also be reclaimed by nature" (Spiropoulou 2010: 108).

Repetition and Confrontation with the Monster

From this point in the cinematic flow, Price and the three survivors understand that in order to remain alive moving was essential. Hence, they drop to the ocean floor and a re-entering of confined passageways to "ova points" is repeated twice more until gaining access to the final control room. As noted previously, the movement through confined spaces seems to be equated with the metaphor of ejaculation. However, Murat Aydemir believes that any repetition of ejaculation metaphors in narrative is "recessive and regressive" (2007: 138) and Eric Savoy (2002: 185), adds that repetition reveals a "death drive and traumatic unbinding of normal logic."

Hence, the drop of water that fell on Price commencing the saltwater expulsions, initiated ongoing figurative attempts by "the Eternal Nature, the eternal feminine" (O'Regan 2002: 52) to reveal to Price that she was the point of choice. Just as the site represented both "the east gate of birth and closer of wombs" (Neumann 2015: 170), she could counter the penetration of Tian by being absorbed into the Great Mother and fight back or stay silent and enable the masculine taint of identity that had begun seeping into her sense of self. Confrontation with the monsters mirrored what she needed to become for the survival of others.

In the final scenes of having reached the command centre with creatures in pursuit, Price finds the escape tubes, but only two are active. It is then that Price makes the decision to save the others through the sacrifice of herself. Her other companions exit and on a screen she sees the monsters chasing the pods to the surface. A large monster appears at porthole window and seems to be reaching out to her. It is then she makes the decision to push the nuclear charge destroying the mining infrastructure, the pursuing monsters and herself. An act that sees her express the "willingness to descend into the male psyche, and to accept the dark" (Bly 1990: 6).

Part III

Masculinities and The Deep

Carl Wilson

Aquaman Volume 1: The Trench (Geoff Johns, Ivan Reis and Joe Prado, 2011–2012)

Aquaman walks into a seaside restaurant; he orders seafood and reassures horrified onlookers that he does not talk to fish as they believe, but rather he telepathically pushes them into action. An enthusiastic self-styled blogger sits uninvited at the table and questions Aquaman's choice of shiny orange outfit, his friendship with animals and his complex Atlantean heritage. If this sounds like the windup for a joke, the blogger makes it clear: "How's it feel to be a punchline? How's it feel to be a laughingstock? How's it feel to be nobody's favorite super-hero?" (Johns 2012: 15).

The restaurant exchange above takes place in issue #1 of the first six-part *Aquaman* story arc of The New 52, which was later published as *Aquaman Volume 1: The Trench* (2012) (see Figure 22). DC Comics' *The New 52* was a relaunch across their entire line of superhero comics. Running from 2011 (after the universe shifting actions of Flashpoint created the Earth Prime universe), each comic series in the line was either cancelled, made anew, or wound back to issue #1. Although this was then reset again after fifty-two issues (following the Convergence and Rebirth events of 2015 and 2016), it was a clear opportunity at the time for characters to be reshaped and redefined in earnest.

While Aquaman may appear to be something of a modern-day joke due to his sea-based connections, considering the range of monsters (King Shark, Abe Sapien, Swamp Thing), gods (Poseidon, Neptune, Triton) and humans (Riptide, Aspen Matthews), he is far from the only aquatic comic character, or even the first superhero of his type. Created by Bill Everett for Timely Comics as a counterpoint to the fire-based Human Torch, Namor the Sub-Mariner appeared in *Marvel Comics* #1 (1939). Namor is half Atlantean, of

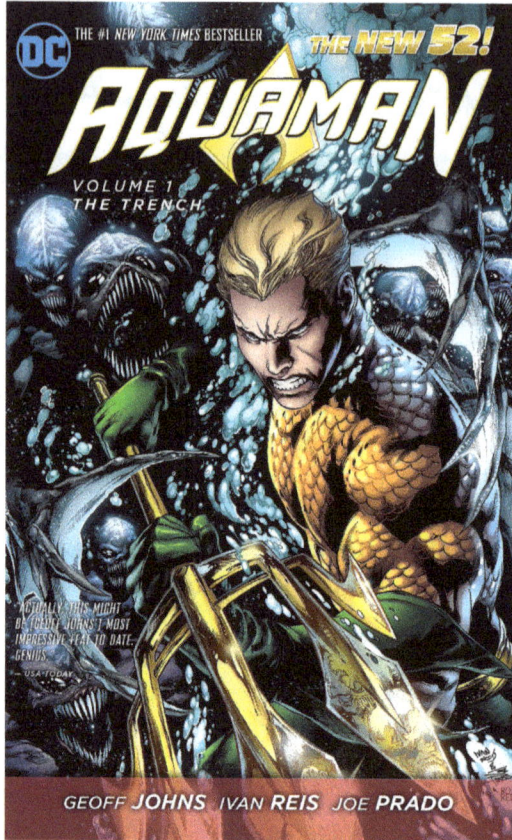

Figure 22. A new Aquaman fights ancient monsters from the Trench (from the cover of *Aquaman Volume 1: The Trench*). Copyright DC Comics reproduced under Fair Use legislation.

royal blood, lives underwater and can be seen fighting Nazis on the deck of a U-Boat in his first cover appearance of *Marvel Mystery Comics #4* (1941). Aquaman was created by DC editor Mort Weisinger, who wrote Aquaman's first story, illustrated by Paul Norris, which debuted soon after Namor, in *More Fun Comics #73* (1941). Aquaman also lives underwater and can first be seen fighting Nazis on the deck of a U-Boat, nine months after Namor. Aquaman's origin story initially posits that his father discovered Atlantis and "by training and a hundred scientific secrets" taught his son to live underwater (Weisinger 1941: 33). In *Adventure Comics #260* (1959), the narrative shifts into him being called Arthur Curry, the son of an Atlantean princess and imbued with the "power to live underwater [...] communicate with sea creatures [and] perform great water feats" (Bernstein 1959: 20). The ways in which comic publishers have directly influenced the creation of new characters from their competitors is well documented (see Morris 2015: 11), but there is usually a crucial point of differentiation. Here, Namor is driven by his anti-heroism, capable of siding with the Avengers or Doctor Doom and Magneto depending on his own objectives, whereas Aquaman became synonymous with the saccharine, selfless heroics of his Gold and Silver Age antics.

There are three key periods in Aquaman's development, leading up to the reflexive moment in *The Trench*: 1941–59 (parallel to the Golden and Silver Age of comics); 1960–85 (Silver and Bronze Age); 1986–2011 (Modern Age). With each accretion of history, Aquaman's relationship with the water and his own identity shifts.

In the first twenty years, Aquaman would ride sea-turtles or porpoises while shouting out jolly, water-based quips such as "I've oceans of love for you boys," (Weisinger 1941: 35) or he would be assisted by "sea friends" such as Ark the seal and Octy or Topo the octopus. These stories are largely uncomplicated adventures; as Brian Cronin explains: "He was just Aquaman 24/7. In fact, we never got a NAME for him besides 'Aquaman.' He was just some guy in the water who helped save the day every month" (Cronin 2018). Grant Morrison offers that "[b]ecause of their status as backup strips in *Adventure Comics*, second stringers like [...] Aquaman weathered the storm" of the Golden Age ending (Morrison 2011: 53). With the influence of the Silver Age of comics, Aquaman was promoted in stature, being given a regular name (not a secret identity) and then a plot-complicating weakness in *Adventure Comics #256* (1958): he must always return to water, or his powers would fade with fatal repercussions. Aquaman's assets at this point then were also the first steps in tethering and refocusing his sea-based character in a different direction, away from humorous romps and derived from punitive measures.

While Aquaman's first period was consistent in establishing his gleeful persona in comics, the following period (1960–85) started to expand and mature Aquaman's portrayal and appeal. Aquaman was one of the seven founding members of the Justice League of America in *The Brave and the Bold #28* (1960). This was then followed by his own self-titled comic in 1962, which ran for sixty-three issues until 1978. As the newly anointed King of Atlantis, Aquaman married Queen Mera in 1964 (*Aquaman #18*): which was a first on the page for the super heroes of DC Comics (later repeated in 2020, post-The New 52 and a new wedding ban). An Aquababy followed extraordinarily soon after (*Aquaman #23*). As Aquaman's sea life was being fleshed out, this chapter also saw an increased sophistication of Aquaman's enemies, such as the introduction of Ocean Master, Arthur's half-brother Orm, in 1966 (*Aquaman #29*), then Black Manta (*Aquaman #35*) in 1967. This period of comic book prosperity, in which the sea became redefined as a domestic space, an imperial space, and one of heightened threats and multi-issue adventures, culminated in a series of Filmnation cartoons, starting with *The Superman-Aquaman Hour of Adventure* in 1967, with footage stripped and recycled into *Aquaman* in 1968, followed by Hanna-Barbera's *Super Friends* in 1973.

However, Aquaman's ascendancy was akin to the bends with his increased success eventually compounding and limiting his cultural depiction. In the same way that Batman's trajectory was irrevocably altered by the eponymous 1960s live-action television series (1966–8), the Aquaman cartoons leant heavily on tropes from his first period, with quips, jokes and amusing nautical scenarios coming to define his character. The comics meanwhile were pulling into more serious waters, focusing on environmental disaster issues. Underscoring this prioritisation of theme over characterisation, when the *Aquaman* comic was abruptly cancelled in 1971, series writer Steve Skeates continued his storyline under Warren Publishing with new Atlantean hero, Targo, then with comic rival Namor in 1972 (*Namor #72*), with a severity far from the storylines of the Aquaman cartoons (see Cronin 2019).

When *Aquaman* was revived in 1977, the hero was given a subversive trajectory, demonstrative to the public of how the comic book Aquaman had also matured since the lingering cartoon chimera period. The plot took up the thread of *Adventure Comics #452* where Black Manta now brought to the fore the theme of race with no half-man/half-Atlantean subtlety: "[S]ince blacks have been suppressed for so long on the surface, they fight well for a chance to be 'masters' below!" (Michelinie 1977: 10). Black Manta also compels Aquaman to engage in gladiatorial combat with his ward, Aqualad, shattering a bond that had endured since *Adventure Comics #269* (1960), and is responsible for the drowning of Aquababy. This latter point leads to a significant amount of post-traumatic grief with Mera and Aquaman eventually separating, an act that immediately gives Aquaman the impetus to disband the Justice League in *Justice League of America Annual* #2 (1984) due to their own parallel lack of commitment to broader issues beyond their own introspective concerns.

The Modern Age is the point at which the multi-reboot comic iterations and transmedia versions of Aquaman, all prompted by these earlier divisions, accelerate, and intermittently collide spectacularly. Following the Crisis on Infinite Earths event that reshaped the DC Comics universes in 1984 and the publication of Frank Miller's *The Dark Knight Returns* (1986), the comic industry would turn towards even darker storylines for their heroes (see Brooker 2000: 172). Superman would die in 1992, but Aquaman had already lost his family and sense of heroic purpose. As such, Aquaman's sputtering renaissance was based on exploring the depths of his heritage. There is the four-part, blue-suited miniseries that ran in 1986; the ancient Atlantean bloodline of Aquaman explored in seven issues of *The Atlantis Chronicles* in 1990; and *Aquaman: Time and Tide*, which re-established Aquaman's origins in four parts across 1993. He was now also known as Orin, with a wizard for a father (until The New 52 changed things back). With momentum behind it, *Aquaman* was relaunched in 1994, running for seventy-seven issues until 2001. This Aquaman had long hair and a beard (which Namor had also recently adopted beforehand), ditched his goldfish top for body-armour and had a harpoon for a hand after an incident with piranhas (see Figure 20). The subsequent 2003–7, fifty-seven issue run of *Aquaman* changed King Arthur's appendage for a magic hand made of water, appropriately gifted by the Lady of the Lake, demonstrating how the The Modern Age Aquaman managed to iterate and innovate in smaller cycles while looking at the wider ripples of water-based myths and fantasy tropes from the past. As such, this gave way to more versions of Aquaman where he is seen to be of a status rivalling the mythological god power of Wonder Woman and the warrior soldiers of Themyscira, such as in the DC Comics miniseries, *DC: The New Frontier* (2004).

Yet, before we get to *The Trench* and where it fits at the end of this lineage and as the start of something new, it is critical that we also consider Aquaman outside of his original domain in the Modern Age. While the comics diverted

Figure 23. "Do you think it's too much?" (*Aquaman #0*, 1994). Copyright DC Comics reproduced under Fair Use legislation.

from the spectre of *Super friends* and company, his cultural image since this time has been less favourably depicted. For every TV pilot spinoff from Aquaman's appearance in *Smallville* (2006) that fails to be picked up, or the poorly reviewed *Aquaman: Battle for Atlantis* (2003) video game, there are more incisive variations that uniformly parody or mock the hero.

Animated series, *South Park* (1997–present), *Family Guy* (1999–present), *SpongeBob SquarePants* (1999–present) and *Robot Chicken* (2005–present) all lampoon Aquaman for his diminished appeal and limited water-based capabilities. Television show *The Big Bang Theory* (2007–19) also had Aquaman as a reoccurring joke with his costume being something of a cosplay short-straw, as is the presentation of "The Deep" in Garth Ennis and Darick Robertson's comic, *The Boys* (2006–12) where he wears an absurd deep-sea diving helmet that he can't remove due to a fabricated ancient Atlantean curse. In the television show adaptation (2019–present), The Deep has an identity crisis and breakdown stemming from the practicality of his atypical powers and abnormal physiology. Attempts at rescuing a dolphin and riding a whale both end with gruesome animal fatalities, underscoring his removal from parodic Justice League equivalent, The Seven.

DC Comics also promoted Aquaman in various animated features of their own, primarily as reoccurring guest appearances within the television shows comprising the DC Animated Universe. These Aquamen frequently match Bruce Timm's stylistic design of the shows, but also loosely reflect his character development in the comics. Of note here is the Aquaman of *Batman: The Brave and the Bold* (2008–11), who has the character history of the middle-period but the over-amplified sunny disposition of the first-era Aquaman. This parody figure has the catchphrase "Outrageous!" and reflexively leans heavily into the kind of sea-based tropes that his colleagues shun him for and most iterations since the 1960s have sought to dispel.

Over seventy years, Aquaman had not only become a "laughingstock" but, by way of counterpoint, he was also presented as embattled by troubles and irritably temperamental; the middle ground had been lost. The New 52 version of Aquaman was an opportunity to clear out these extremes and to get to the core of the character.

As Mera pointedly states towards the end of *The Trench*: "I don't know how Arthur can put up with this constant barrage of unappreciated and unoriginal comments," quickly followed by: "The absence of water is not my weakness. It's all of yours" (Johns 2012: 120). In this reframing of Aquaman's dynamics, the misconceptions are revealed to lie with the ordinary public that have constructed the Aquaman persona; this Aquaman has only tolerated the mocking that surrounds him. Because of the positive aspects of being birthed from the crushing pressures of the deep sea, Aquaman is bullet-proof and capable of great feats of strength. By leaning into this constructive approach, where his water-sourced powers are an asset, for the first time since 1958 Aquaman no longer required perpetual hydration to survive.

In Ivan Reis' annotated sketches for The New 52, Aquaman is younger than the grizzled veteran that preceded him, so there is no beard and less opportunities to have lost an extremity; as Aquaman explains to Mera, while discussing his upbringing: "I did a lot of things I don't need to do again" (Johns 2012: 33). Their design is also quoted as referencing *Brightest Day* (2010–11), a comic series that Geoff Johns and Ivan Reis had worked on previously (Johns 2012: 142–3). As with all the DC Comic character iterations over time, The New 52 is not entirely new then, but it is recombined and refreshed.

The Trench opens with the police being baffled by Aquaman's crime-fighting presence on land, then the narrative turns to the water, not as a friendly place for seahorse riding, but as a Lovecraftian nightmare from which creatures that "you can't even begin to imagine" emerge (Johns 2012: 50). A military man offers "[w]e'll put you in a good light for once," (Johns 2012: 59) but in this horror story about the food chain, Aquaman is no longer ineffectual and ridiculous, and neither is the sea; after diplomacy with the alien queen fails, Aquaman fights the marine monsters not with water and jokes, but with lava and the threat of genocide.

Figure 24. Jason Momoa embodies the same but different Aquaman as those that have come before him (promotional image for the *Aquaman* movie, directed by James Wan (Warner Bros., 2018)).

With The New 52, Aquaman had both his dignity and classic associations restored, which has paved the way towards his feature film appearances in the DC Extended Universe, appearing first in *Batman v Superman: Dawn of Justice* (2016) then *Justice League* (2017) and his own movie with *Aquaman* (2018) (see Figure 24). Given that Momoa played Conan the Barbarian onscreen in 2011, the appearance of a bearded Jason Momoa matches the Aquaman as Underwater Conan brief aimed for in the pre-52 comics and media, as Aquaman fights creatures adapted from *The Trench* with costumes designed on those derived from *Brightest Day*. This ability to update Aquaman with an understanding of the hereditary strictures of historical and cultural constructs emanates directly from *The Trench*, making space for cinematic spectacle but also further nuance and commentary on the character. As Lyn Dickens points out, "*Aquaman* is notable for its explicit links between the fantasies of hybridity and the lived experience of mixed-race people" and this is embodied in the casting of Jason Momoa (Dickens 2019). Aquaman's history of not being like the other superheroes has finally stopped being seen as a hindrance and is fully celebrated.

Changing Masculinities

Gerard Gibson

The Creature from the Black Lagoon (Jack Arnold, 1954)

Finding its origins in the ancient legends of mermen, merrows and water-dwelling sprites, Jack Arnold's *Creature from the Black Lagoon* (1954), takes these languishing archetypes and re-imagines them in a form that has become a lasting popular culture icon. The physiological anthropo-morphism of the Creature, and in particular its contested masculinity and humanity, offer a unique vantage point from which to explore the pressures and anxieties affecting the white American male in the 1950s.

Taking the form of an adventure into the unknown, the film follows a straightforward plot, owing much to Conan Doyle's *The Lost World* (1912). Discovering a fragmentary fossil that suggests a previously unknown animal – an evolutionary missing link bridging water-dwelling creatures and bipedal land animals – Drs Maia (Antonio Moreno), Reed (Richard Carlson) and Williams (Richard Denning) search a remote region of the Amazon basin for further evidence. Accompanied by Kay Lawrence (Julie Adams), Reed's girlfriend and assistant, and Lucas (Nestor Paiva), skipper of the Rita, they moor in an unexplored lagoon associated with local legends of a river dwelling monster. When a creature, the Gill-man (see Figure 25), is drawn to Kay as she swims, a series of events are set in motion, leading to encounters which will change all their lives.

Emerging from the weeds and deep shadows of the Black Lagoon, the Gill-man casts a long shadow across the history of horror, while also perhaps offering a uniquely valuable window into white American masculinity in crisis in the 1950s. These gendered anxieties and conflicts first emerged during the film's production stage. When initial costume tests foundered, Milicent Patrick, an ex-actress and then part of Universal's Make-up Department, offered a

Figure 25. The Gill-man. *The Creature from the Black Lagoon,* directed by Jack Arnold (Universal Pictures, 1954).

creature design which effectively saved the production, one that has become a lasting cultural icon of horror. Bud Westmore, highly competitive head of her department, growing jealous of the attention his photogenic protégé received from the studio and the public, had her dismissed out of hand and took the screen credit for himself. Patrick's significant contribution was only acknow-ledged decades later, in the years after Westmore's death (Skotak 1978: 16–20). Patrick's Gill-man, the last of the great Universal Monsters, is an antagonised and brutalised figure at the centre of a trilogy of films, which time and again resonate powerfully with the forces and experiences fracturing 1950s white American masculinity.

In Arnold's film, the Gill-man first confronts those who invade his remote realm, searching for fossils in the nearby riverbanks, but then evades the Americans who visit the lagoon. While attempting to elude divers who seek to capture him, without provocation, the Creature is harpooned in the back by Dr William's, who sees it as a game trophy, something to be killed and possessed. When the Gill-man escapes, Williams and the others poison the lagoon with Rotenone, an effective pesticide killing the fish and disorienting the amphibian. They then follow him into the caves and grottoes around the lagoon to trap him. Repeatedly across the original film and its sequels, *Revenge of the Creature* (Arnold, 1955) and *The Creature Walks Among Us* (Sherwood, 1956), the Gill-man is harpooned, poisoned, shot, imprisoned, experimented on, tortured and even set ablaze. His home and habitat are repeatedly invaded and rendered toxic. He is assailed from every side.

The Gill-man emerged into a decade of dramatic, even bewildering change for the white American male; a period, Gilbert (2005: 2) argues, more pre-occupied than any other with what American masculinity was supposed to mean and how it should be expressed. The male role was highly contested, challenged by a number of complex forces. Perhaps chief amongst these was the unrecognised and unresolved trauma resulting from participation in significant military action. The motifs of physical assault and injury borne by the Gill-man may have resonated strongly with many serving males, appearing as

Figure 26. Shot in the back. *The Creature from the Black Lagoon*, directed by Jack Arnold (Universal Pictures, 1954).

he did on the screen in February 1954, just six scant months after the close of a particularly problematic and damaging war.

American men of all colours had put their bodies and lives on the line in the Second World War. Some of the veterans and enlisted men who faced the horrors of Nazi Europe and the terrible war in the Pacific, were barely home before the conflict in Korea required an intervention by United Nations forces and the US Army in June 1950. The war-machine of the United States was, because of the Korean peninsula's uniquely challenging geography, unable to deploy its heavily mechanised might, and armoured vehicles were left to function in largely supporting roles. This arduous terrain offered advantages to the more lightly armoured Chinese and North Korean forces (Tucker-Jones 2012), who exploited this opportunity. Despite an undeniable American superiority in the air, the ground fighting was intense and bloody. During this brief but ferocious conflict, American forces suffered twice as many loses per annum than would later be suffered in the more prolonged Vietnam War (Department of Veterans Affairs 2021). The brunt of this terrible attrition fell upon the ground troops. This intense, savage combat ended in stalemate and an ignominious armistice in July 1953. Through their involvement in what Fehrenbach (1963: 655), refers to as the most forgotten war in American history, combat veterans, many no doubt suffering what we now recognise as Post-Traumatic Stress Disorder, returned home not to recognition or even acknowledgement, but to shame and obscurity. Many years would pass, scabbing over these deep wounds before this failed campaign would even be recognised or referred to as a war (ibid.).

Nor was white American masculinity uncontested in the home they returned to. With the Cold War at its height, fears for national security soared. Vigilance against threats from outside and within the United States may have lent significantly greater pressure to the always extant coercive social forces which demand conformity. Between 1950 and 1954, fears rose about activities that were "Un-American." Senator Joe McCarthy's witch-trials reached their frenzied zenith in this period, exerting incredibly powerful intimidatory forces which offered security against communism by encouraging Americans "to see dissent as disloyalty" (Murrow 1954), and carry out surveillance on each other. Such protection was afforded at the cost of lost civil liberties and perhaps unparalleled socio-political pressure to comply in what Preston (2012: 31) refers to as a "sinister conformity." Fears surrounding communism were further conflated with anxieties about homosexuality and politicised in "The Lavender Scare" which ran across the 1950s and 1960s (Charles 2012). Even heterosexual masculinity was in a state of anxiety, divided against itself, with conflicts socially playing out in the press between male muscularity and masculine intellectualism, the aggressive and the sensitive, a struggle between the body and the brain (Gilbert 2005: 3).

These socio-cultural motifs are very evident in Arnold's film, particularly in the exchanges between Dr Williams and Dr Reed. Williams sees his pursuit of knowledge as a means to acquire power, acclaim and wealth, as is suggested by the dialogue early in the film between Kay and Dr Reed about the endowments that Williams' publicity seeking has brought to their institute, securing their careers. The economic precarity of academia has left them feeling beholden to Williams. Williams' adversarial muscularity is also evident in his gleeful hunting of the evasive Gill-man, even bludgeoning the unconscious Creature when he eventually finds it. Reed, on the other hand, is thoughtful and sees knowledge and understanding as the important goal. He is, however, undermined in his efforts by his economic reliance on, and subordination to Williams.

Whilst much is written on the relationship between the Gill-man and Kay and the implications and meaning of their underwater pas de deux (Banner 2008: 5; Verevis 2012), fresh examination of the film suggests that the Creature shows little real interest in Kay beyond their shared swim. There are no other scenes showing any fixation on Kay, no shot-reverse shot associations, no evidence of the Gill-man looking particularly at her or following her, such as is demonstrated in scenes in the film's sequel *Revenge of the Creature*. These clearly indicate a sustained interest in Helen Dobson (Lori Nelson), that film's female lead, and ultimately reveal an implied sexual threat. In *Black Lagoon*, however, when the sensuality of their shared underwater ballet has finished, the Creature seems to even fear touching Kay. Once out of the water none of the male characters act or speak in a way that suggests they consider Kay to be at particular risk or of special interest to the Gill-man. Nor does she show any dismay. Though the original script bore lines spelling out the Creature's interest in Kay, these were cut specifically at the behest of MPAA censor (Breen, 1953), and the motif diminished. After Kay's swim no other interest in her on the part of the Creature is referred to or implied in any surviving dialogue or footage. Any relationship between the Gill-man and Kay has thus received little actual narrative development. Perhaps some of the attention focused on a theme which barely survives in the finished version springs from the numerous paratextual publicity shots of the Gill-man menacing Adams in her specially designed swimsuit. These scenes are not in the film and not part of the story but suggest a more substantial and sensual relationship than is actually evident in the exhibited cut. Simply put, in the final film the Gill-man does not seem to focus much on Kay to any extent once she is out of the water. When he finally sweeps Kay off the deck of the Rita, he might just as easily be grasping the only human present on deck who has not attacked, harpooned, poisoned, burned, or shot at him, as much as scooping up any target of frustrated amorous intent.

Whilst suggestions of non-normative interspecies romance may have upset the MPAA censors, it still had ostensibly heterosexual dimensions. Little, however, has been said about what is the real core dynamic of the film's drama – the complex, contested, even intense relationships between the three central male characters; the sensitive and respectful Dr Reed, the brash and ambitious Dr Williams and the Gill-man himself. To Williams, the Creature is merely an object, a scientific and evolutionary curiosity which he must acquire, dead or alive, to further his aggressive career ambition. Dr Reed, on the other hand seeks knowledge rather than acquisition, wanting to study and photograph the Gill-man in his environment. In their search of the lagoon, Williams deploys a multi-pronged, even vaguely phallic-looking, gas-powered harpoon, whereas Reed is armed only with a boxy underwater camera.

The dynamic between Williams and Reed has other more overt political overtones. When Reed, reluctant to aggressively hunt the Creature, challenges Williams' intent, Williams unflinchingly reminds Reed that they are not equals, and that, though they are far from home, Reed is still an employee. This exchange makes Reed subordinate, both "feminising" him and at the same time uncovering another tension troubling American masculinity at this time; the rise of corporatism and the loss of individual agency in labour. The rise of the mega-corporations in the post-war economic boom emasculated many American men, effectively turning the grandsons of pioneers into paper-shuffling pencil-pushers. Independence and self-determination were a

Figure 27. White American Male pulchritude in crisis as displayed in the face-off between Dr Reed (Richard Carlson) [left] and Dr Williams (Richard Denning) [right]. *The Creature form the Black Lagoon*, directed by Jack Arnold (Universal Pictures, 1954).

thing of the past, as yesterday's self-made men had become a nation of do-as-you're-told employees (Kimmel 2006: 158) and (Joyce 2011: 48).

Each of these themes – acquisition versus knowledge, violence versus non-aggression, muscularity versus intellect and the exertion of academic and economic hierarchies – emerge in the character interaction and the dialogue but are equally rendered visible on screen in the athletic yet vulnerable corporeality of the Gill-man and his human counterparts. As if to exemplify this, in the scene where the Creature finally catches Kay in his arms, she is fully clad, yet Drs Reed and Williams are dressed in just trunks and belts; American male pulchritude laid bare and on show (see Figure 28). While Kay has just one aquatic sequence, these active male bodies are repeatedly on display, questing, hunting, swimming, minimally clad in clinging apparel, moving across the screen often in long scenes accompanied only by music, the twists and turns of their bodies presented for the camera. The Gill-man moves fluidly in the water, but lurches on land, deliberately played in this way by different actors, the role divided due to production constraints. Ricou Browning performed the swimming scenes shot in Wakulla Springs, Florida and Ben Chapman wore the suit on the Universal backlot in Los Angeles. The swimming movements of Reed and Williams seem forced when compared to the Creature's graceful, near-languid underwater choreography.

Arnold's highly successful film was released in February 1954, and debuted in some theatres in March across America accompanied by newsreel footage of the disastrously contaminative Castle Bravo hydrogen bomb test. With this accidental juxtaposition on the cinema screen, the merely monstrous pales into insignificance in the company of the truly horrific. *The Creature from the Black Lagoon*, touched a nerve in the American imagination and its box office success spawned two direct sequels. These too, each reiterate and extend the same motifs which suggest so much about white American masculinity in crisis.

The archetype of the Gill-man, the power and imaginative artistry of Milicent Patrick's design, the skill of suit fabricators Chris Mueller and Jack Kevan, the performances of Ricou Browning in the water and Ben Chapman, Tom Hennesy and Dan Megowan on land, despite receiving no recognition at the time, have all allowed the Creature to reverberate through the history of fantastic cinema. His descendants are varied and appear across a myriad of media, but they all still carry forward the aesthetics and ideas so well laid out in the original trilogy. The Gill-man appears as one of the cardinal monsters in Dekker's *The Monster Squad* (1987), he has profoundly influenced Mignola's Abe Sapien character from the *Hellboy* comics, graphic novels, games, animations and films. Peeter's *Humanoids from the Deep* (1980), makes explicit the

Figure 28. Spawn of the Gill-man, Mythrol (Horatio Sanz) from *The Mandalorian*, directed by Carl Weathers (Lucasfilm, 2021).

unspoken sexual threat of *Revenge of the Creature*. This relationship is given more erotic and romantic dimensions in del Toro's Oscar-winning *The Shape of Water* (2017), whose awards perhaps also finally acknowledge, in part, the Creature's imaginative and creative legacy. Indeed, the Gill-man has become such a stable of the fantasy canon that it even appears in Star Wars universe, in the shifty form of the Mythrol (see Figure 24). When designing for Favreau's *The Mandalorian* (2019–present), the brief to the conceptual design team from the director for this particular character, was specifically to design a Gill-man (Szostak 2020: 40), and some of the design development resembles the original Creature.

The Gill-man, born of ancient legends and adventure, offers a mirror through which humanity, and in particular, ideas of contested masculinity can be examined, explored and evaluated. It is no coincidence that in Sherwood's *The Creature Walks Among Us* (1956), that after once again being burned, life-saving surgery reveals beneath the Gill-man's glistening scales an all too human skin. These films invite us to consider what it means to be human; to be a white American male in a particular time of change, confusion and crisis and as these films unfold, they encourage us to question just who might really be the monster.

Brigid Cherry

The Shadow Over Innsmouth (H. P. Lovecraft, 1936)

Sea monsters loom large in H. P. Lovecraft's fiction: arguably they are most notable in *The Shadow Over Innsmouth* novella ([1936] 2015). In recent years, *Innsmouth* has inspired a number of comic books, adding to the wide range of comics and graphic novels contributing to the shared story world of the Lovecraft mythos (see Sederholm and Weinstock 2016: 12–15). Among the comics drawing on *Innsmouth* as an antecedent text (McFarlane 2007: 19) are *Hellboy: Seed of Destruction* (Mignola 1994), *Only the End of the World Again* (Gaiman and Russell 2000) and *The Squidder* (Templesmith 2015). Two further examples – *The Shadow Over Innsmouth* (Marz and Rodriguez 2014) and *Neonomicon* (Moore and Burrows 2010–11) – are discussed in this chapter in relation to significant concepts of adaptation studies (a vital approach given the breadth of the Lovecraft mythos in contemporary culture). In terms of remediation for the comic book medium (Bolter and Grusin 1999: 273), the Deep Ones – ocean-dwelling creatures that mate with the humans of Innsmouth giving the townsfolk fish and gold artefacts in return – form a key intertext in both these comic books that engages with current social and environmental concerns. Both also function as Lovecraftian metatexts (Keller 2008: 189), constructing commentaries on Lovecraft's writing and themes in relation to his current-day popularity and issues with his xenophobia.

Remediating Innsmouth

The Shadow Over Innsmouth is a punning title alluding to the contemporaneous pulp character The Shadow (popular in the 1930s when *Innsmouth* was written) alongside the Lovecraft connection. The comic thus works as an intertext of The Shadow and the Lovecraft mythos. Signposting this, the cover is dominated by The Shadow (Lamont Cranston) wearing his trademark black cloak, wide-brimmed hat and red scarf covering his lower face above a detailed illustration of the New England seaport. The promise of Lovecraftian sea creatures is communicated via writhing cephalopod tentacles emerging threateningly behind him. The comic book opens with a splash page, a single panel with an establishing high-angle, tilted shot of the harbour with jetties and boats, and the Innsmouth Inn central in the frame. A speech bubble emerges from the inn with the line "I'm still not quite sure *what* happened ..." (Marz and Rodriguez 2014). The angle of the horizon and seaport buildings invoke a sense of eldritch horror and the liminality of the inn in relation to the sea. The dialogue – Margo (Cranston's assistant) recounting an attack – suggests mystery. The story that follows is framed around Cranston's investigation of events in the town, and the plot follows the conventions of the detective story, investigating and apprehending a gang of criminals. This generic hybridity draws attention to the hybrids (the fish-human creatures that result from mating with the Deep Ones) of the mythos.

Accordingly, *The Shadow* remediates eldritch horror in the service of a Shadow plot. The narrative pivots on the idea proposed in *Innsmouth* that events were "[a] contagious nightmare hallucination," (Lovecraft 2015: 866) though falling definitively on the side of rational explanation in clear opposition to the Lovecraftian supranatural. Nonetheless, the connections to *Innsmouth* as antecedent text are made explicit through two double splash pages that emphasise connections to the sea. The first stands out by being drawn in monochrome shades of a sickly green seaweed colour that contrasts with the rich warm tones – deep reds, browns and yellows predominate – of the surrounding pages. Each inset panel, depicting the hoard of menacing fish-men that advance upon Margo (see Figure 29), is edged with a thick black frame that give the pages a heavy, muted feel, and – echoing the cover – entwined by

Figure 29. Bootleggers dressed as Deep Ones in *The Shadow Over Innsmouth*, © Advance Magazine Publishers Inc., reproduced under Fair Use legislation.

grasping, nodular cephalopod tentacles that convey the horrors from the deep of the antecedent text. The second tells the story of Captain Obed Marsh that is part of the folklore of Innsmouth, thus borrowing directly from *Innsmouth*. The illustration echoes the colours of the previous splash page with a green tinge, but a sepia colouring is also used, conveying the sense that these are both historical events and – via sepia's origin as squid ink – monstrous. The inset panels depict Marsh learning of the Deep Ones and importing the Esoteric Order of Dagon to Innsmouth. Father Dagon, Mother Hydra and Cthulhu are all referenced, with the panels depicting human-Deep One sex and fish-frog hybrid creatures. This meets the expectations of any reader familiar with *Innsmouth*, but the promise of the cover and these splash pages is disrupted. Despite Margo's encounter with the fish-men, Marsh's history is passed off as a "ridiculous fable" by Cranston and even the storyteller, Zadok Allen, says it is just a "story people around here tell," a local legend that he does not claim is true. This is a very different Allen to the one described in *Innsmouth*: a destitute, nonagenarian drunkard. *The Shadow*'s remediation of Allen fills the role of *Innsmouth*'s Allen in regaling Cranston and Margo with his fanciful tale which matches point by point the one that Allen tells in the novella, but there the parallel to *Innsmouth* ends.

The Shadow instead goes along with the newspaper cover-up suggested in the novella, though in this case the "sea monsters" really are bootleggers. In fact, when Margo investigates further and is attacked by a fishman, she fights back and claws at his face, ripping off his mask in the process (it is, of course, Allen). There are no fish-human hybrids or Deep Ones here except in the local lore that seeds the imagination of The Shadow comic's Lovecraft. This is revealed when *The Shadow* adds a metatextual coda referencing Lovecraft himself. The final panels depict the publican at the inn recounting the destruction of the bootlegger's warehouse: "I heard they pulled somebody out of the warehouse *alive*," he says. "They asked him what happened and you know what he said? He said, 'The Shadow' and then he *died*" (Marz and Rodriguez 2014). The customer is seen first through a beer glass, then half in shadow and then a close-up of his hand, writing, but there is a clear resemblance to any reader who recognises Lovecraft. And this is indeed confirmed when the publican asks: "What do you think about *that*, Howard?" Lovecraft is obviously inspired, the panel depicting a clear shot of his face, pen poised, as he muses on the publican's report: "A shadow? In Innsmouth? I like that." In the final panel, he resumes his writing. Here the inspiration for the fictional Deep Ones is a local legend which connects Innsmouth with the sea but has (seemingly) little basis in reality.

Intuiting the Deep Ones

In *The Shadow* narrative, the Deep Ones thus predate Lovecraft's fiction, if only as local folklore exploited by bootleggers to scare off observers. They are not diegetically "real," and Lovecraft's writing is simply fiction inspired by an implausible old seaman's tale. Alan Moore's remediation of the Cthulhu Mythos – taking place across *The Courtyard* (1994), *Neonomicon*, and *Providence* (2015–17) – similarly suggests that Lovecraft took inspiration from pre-existing knowledge of the Deep Ones, or was perhaps "intuiting the truth" of the world, in his writing. But in contrast with *The Shadow*, the Deep Ones, along with the Great Old Ones and their realm, exist physically in the *Neonomicon* story world.

In a similar generic hybridity as *The Shadow*, *Neonomicon* is also framed as a detective story. Recognisable details from Lovecraft's fiction form clues in the case being investigated by FBI Agent Merril Brears, something of a Lovecraft scholar. Brears initially assumes the criminals, like the bootleggers, are using the Lovecraft connections as cover for their activities. Her partner Gordon Lamper even expresses *The Shadow* defence: "They use this H. P. Lovecraft gimmick to scare folk off and they've been doing this since the 'twenties'" (Moore and Burrows 2010–11). And Brears goes on to suggest that since nobody knew Lovecraft's work back then, maybe he based his work on their "rackets." She thus proceeds to identifying the clues – the Zothique club in Red Hook, the speaking of the Aklo language, the characters named Randolph Carter and Johnny Carcosa, and bands named Rats in the Malls and The Ulthar Cats – as "a literary in-joke" (this can be read as Moore self-referentially acknowledging his own authorial tropes as well as the antecedent text). *Neonomicon* not only creates a Lovecraftian intertext in this way, but also explicitly engages with representations of Deep One-human procreation that underpin the fear of racial contamination expressed in *Innsmouth* (Newitz 2020: 247–8).

Connections to eldritch sea creatures recur throughout the text. Club Zothique is located in an old waterfront church repurposed for the Order of Dagon, Aklo is described as "talking like he'd got an octopus in his mouth," starfish and ammonite imagery is seen in Carcosa's apartment, "weird starfish dildos" in the cultist's new age/sex shop in Salem, and the cover designs include stylised squid-like icons. The tunnels under the cultist's shop are connected to the ocean, allowing the Deep Ones access to their orgies. These visual and textual details make the sexual aspects of the Deep Ones overt. In *The Shadow*, the bootlegger's costumes reproduced the repugnant features of the Deep Ones, inked in dank greens with opaque white eyes, wide down-turned mouths, finned heads and webbed fingers. They have a male physique (presumably that of the smugglers inside) but, if not exactly sexless, with no genitalia apparent. *Neonomicon*'s Deep One (see Figure 30) on the other hand is extremely sexualised and plays on the sexual implications of human-Deep One interaction. It

Figure 30. The Deep One with Agent Brears from *Neonomicon*, © Alan Moore and Avatar Press, reproduced under Fair Use legislation.

is depicted as a sexualised fantasy of the male body, large and imposing physique, extremely muscular, with erect spines on its back, and with an enormous erect penis. Its extreme sexual prowess is worshipped by the cultists in their orgy, and its capacity for sex is represented by its repeated ejaculations and rape of Breer's multiple times over several days. *Neonomicon*'s Deep Ones are thus a form of sexualised monstrosity, and one which incorporates notions of toxic masculinity. The realistic horror here – rape and sexual violence – also encode the xenophobic if not racist aspects of Lovecraft's writing. The African-American Lamper is subjected to the cultist's casual racism – calling him a "Black boy" and the n-word, comparing him to Nyarlathotep (the dark Outer God who resembles an Egyptian Pharaoh), and referring to his "Black power," before they kill him (Moore and Burrows 2010–11). His body is dragged away as Brears is offered up to the Deep One, the racially problematical myth of Black potency elided by that of the monster's. The pivotal moment in the narrative comes when the Deep One realises Brears is pregnant – and thus mother to the monster, much like the progenitors of the hybrid inhabitants of Innsmouth. This is the point at which he frees her from the cultists' pool, but more importantly, as later revealed, the child is the sleeping Cthulhu and her womb therefore a figurative R'lyeh. When she dreams she is in the drowned city, Carcosa tells her that "R'ylah is in you." Thus, *Neonomicon* directly (and problematically in terms of representations of femininity) connects the ocean from which Deep Ones rise with the water of the womb.

Confronting Lovecraft

The Lovecraftian theme of genetic contamination or impurity is relocated around class-based criminality in *The Shadow* and around rape and racism in *Neonomicon*. The Shadow comic renders the final sentence in the opening paragraph of Lovecraft's novella literal: "Uninquiring souls let this occurrence pass as one of the major clashes in a spasmodic war on liquor" (Lovecraft 2015: 866). It also draws attention to the comment (by the ticket agent to the narrator) in the novella that the people of Innsmouth are "what they call 'white trash' down South – lawless and sly, and full of secret doings" (871). The "monsters" here, the criminal class as opposed to hybrids of human and Deep Ones, are contrasted with Cranston's status as a wealthy metropolitan. Drawing on his comics and pulp fiction background as a cosmopolitan

New Yorker, a wealthy playboy and a famed aviator – Cranston arrives in Innsmouth in his own seaplane, and dressed in a dark business suit and tie, he is in opposition to what de Bruyn describes as "the impoverished New England gentry" (2015: 98) that concern Lovecraft. This recoding of the horror of Innsmouth's sea monsters as a masquerade to cover criminality is in keeping with a story focused on The Shadow, but it nonetheless draws attention to Lovecraft's themes of monstrosity as class impurity, and also encodes the disparity of wealth in contemporary America.

In *Neonomicon*, eldritch horror underlies the outwardly "hip" new age milieu of Salem, just as folklore does Prohibition-era America in *The Shadow*. The orgiastic narrative and gendered subtext of *Neonomicon* addresses Lovecraft's apparent attitudes towards sex. During the sexual assault on Brears' body, her mind ponders Lovecraft's biography: of his mother dressing him as a girl when a child, being raised by prudish aunts, of how he remained clothed when making love to his wife, of how he "had barely got a sexual bone in his whole body" (Moore and Burrows 2010–11). As an uncomfortable depiction of the horror of rape, this can be considered alongside other Moore texts which di Liddo refers to as "sexualised reinterpretations of [...] literature" (di Liddo 2009: 46). The description of Lovecraft is placed in direct opposition to Brears' experience of the Deep One and her own sexual experiences as a recovering sex addict. Significantly, the connection between the female body (Brears' womb) and the ocean (Cthulhu) implies contamination of a different sort to Lovecraft's fear of genetic impurity – namely, the pollution that plagues the world in *Neonomicon* with its cities shielded by anti-pollution domes.

As a postscript, it is worth noting that Megan James' *Innsmouth* (2016– 18) confronts Lovecraft's xenophobic sensibilities in a different way, that is via parody. James states in her foreword that while reading Lovecraft as a child she "loved every fish monster, evil cult and doomed protagonist," but came to realise that the fear of the unknown in his work was fear of anyone who was not "[w]hite, straight, educated, Anglo-Saxon." In response, her Innsmouth is populated by bourgeois townsfolk, a lively mix of hybrids and humans living in multiracial (human-Deep One-hybrid) families and sitting on town committees, for whom the return of the Old Ones is an annoying impediment to their schedule of potlucks and weddings. One of the main characters is Fatima, a hijabi and descendant of the Abdul Alhazred who wrote the Necronomicon.

Figure 31. The Deep One in *Innsmouth*, © Sink/Swim Press, reproduced under Fair Use legislation.

She aids the nerdy Randolph in preventing the End Times along with Dr Herbert and Dr West, here a gay Black couple. Moreover, contrasting sharply with the criminals' costumes in *The Shadow* and the sexualised Deep One in *Neonomicon*, James' Deep One (see Figure 31) is extremely corpulent, dresses in voluminous priestly robes and diadem, and communicates with the cultists via Billy Bass-style mounted fish. James' depiction elides the horrors of the novella (and *The Shadow* which draws on it) *and* the monstrous sexuality of *Neonomicon*, all the while confronting and countering the racist overtones in Lovecraft's work. In common with the other comics discussed here, such adaptations reclaim the Lovecraft mythos, not only by metatextually addressing his racism, classism and asexuality, but by representing Innsmouth, the Deep Ones and humankind's relationship with the ocean in relation to contemporary understanding on these issues.

Kodi Maier

Hans Christian Andersen (1805–1875)

On the rocky shores of Copenhagen, Denmark, a mermaid wistfully gazes out to sea at the passing ships. Not completely human, still bearing the vestiges of her fish tail, the bronze statue is based on Danish writer Hans Christian Andersen's famous 1837 fairy tale, "The Little Mermaid." Across the Atlantic in Orlando, Florida, another mermaid, Ariel, watches as theme-park guests pour into her auditorium to see "Voyage of *The Little Mermaid.*" The live-action performance recreates Disney's classic animated film, *The Little Mermaid* (1989), and features the songs producer and lyricist Howard Ashman wrote for Ariel before his untimely death in 1991. Because the film is itself a retelling of Andersen's fairy tale, the mermaids share a longing for the human world and the impossible love of their handsome human princes, as well as a sense of isolation within their undersea kingdoms for their taboo desires. Wistfulness, longing, loneliness, isolation: these melancholic feelings link both mermaids via queer affect. This queer affect penetrates deeper than their stories, as both tales were inspired by their creators' queer feelings: Andersen's unrequited, passionate affection for Edvard Collin and Howard Ashman's struggles as an openly gay man with AIDS. This chapter thus argues for a genealogy of queer affect, specifically longing and loneliness, beginning with Andersen's feelings as an outsider in his society and concluding with Ariel's desire to be part of the human world.

Impossible Love: Andersen's and His Mermaid's Queer Affect

Queer Affect

Multiple scholars have claimed Andersen as part of the LGBTQ+ community over the years. Sean Griffin writes as though Andersen's gay identity is all
but settled fact, noting he is "now recognized as a homosexual" (2000: 64).
Robert W. Meyers is more tentative, arguing for an "unconscious feminine
identification" (2001: 154) that lent itself to "strong homo-erotic" (158) attachments. Others, noting Andersen's multiple courtships with women like
opera singer Jenny Lind, make the case for Andersen's bisexuality: Gabrielle
Bellot alludes to this when she writes that "Andersen frequently flirted with
men and women alike" (2019). Boze Herrington believes that Andersen's al-
leged bisexuality was closer to biromanticism and that the Danish writer was
somewhere on the asexual spectrum (2019). However, to retroactively claim
Andersen as any part of the LGBTQ+ community is disingenuous and "to
think of Andersen as a queer writer [...] somewhat anachronistically exports
the contemporary meaning of the term" (Hamilton 2009: 246).

We do not know how Andersen contextualised his romantic and platonic
relationships, but we do have ample record of Andersen's deeply queer *feelings*. Multiple scholars (Griffin 2000; Hamilton 2009; Meyers 2001; Zipes
1991) describe Andersen as an outsider due to Danish society's intolerance
for his non-normative desires, and this is true. His affection for his adoptive
brother, Edvard Collin, as well as Charles Alexander, the Grand Duke of Saxe-
Weimar-Eisenach, and Harald Scharff, a ballet dancer for the Royal Danish
Theatre (Meyers 2001: 158), did not conform to heteronormative expectations
by any means. Yet Andersen was queer – that is, strange, melancholic, at odds
with society – for reasons beyond his affections. Bellot describes him as "a bit
supernatural," "a little bit puckish behind the eyes," with "a faintly haunting
melancholy to him" (2019); Jennifer Hamilton calls him a "socially inept outsider on his particular and crooked path" (2009: 244); and Jack Zipes char-
acterises Andersen as "the outsider, the loner" who "never felt himself to be a
fully fledged member of any group" (1991: 72).

Born to a cobbler and a washerwoman, Andersen escaped his rough roots
at seventeen, when Mr Jonas Collin, a civil servant, effectively adopted him so
he could attend grammar school and develop his budding talent as a writer.

However, Andersen "never felt fully accepted by the [Collin] family; he saw himself as a homeless outsider" (Meyers 2001: 156) because "he measured his worth by the standards *they* set" (Zipes 1991: 74, emphasis in text) – standards Andersen invariably failed to reach. We get some sense of the rejection and isolation Andersen felt through his fairy tale, "The Ugly Duckling." Torsten Thomsen notes that "Andersen himself did make a connection between 'The Ugly Duckling' and his personal life – especially his childhood and youth – viewing himself as the duckling."[1] In the story, the duckling is not only rejected by his birth family and exiled from his home, he is also constantly reminded that he will never be fully accepted in the outside world. He is always too ugly, too strange, or too queer for the other animals he encounters. It is only when the ugly duckling meets the "noble," "royal" swans and discovers that he is one of them that he finally finds approval and love (Hersholt 2019 translated from Andersen 1843). Thomsen links "The Ugly Duckling" to a turning point in Andersen's life that occurred in 1838, when "the king decided [...] to grant him lifelong art support through annual payments from the king's personal foundation" (2021). Thus, much like the duckling who found his home among the swans, Andersen finally found a place among the aristocracy. Yet because Andersen was born "a member of the dominated class, [he] could only experience dissociation" (Zipes 1991: 83) in such circles. In other words, though the writer successfully passed amongst the aristocracy, he remained a lonely, melancholic outsider.

It is these persistent feelings – loneliness, melancholia, longing, strangeness – that ultimately queers Andersen. Affect "is conceived as an embodied understanding, or as a sense of feeling that provides an interpretation of the social conditions within which we reside" (Johnson 2015: 122). Because "[h]omosexuality has a long history of abjection, religious condemnation and state persecution," (122) the LGBTQ+ community has long been well acquainted with feelings of rejection, loneliness, strangeness and melancholia, among other negative affects. Heather Love describes queers as "a backward race," (2007: 6) wherein "backwardness" runs a gamut of negative affects and socially undesirable states, including melancholia, loss, failure, impossible love, longing, loneliness, self-hatred, despair and shame (6, 21, 146). While

---

1   Personal correspondence with Torsten Bøgh Thomsen (26 March 2021).

these backward feelings are not unique to the LGBTQ+ community, they are persistent enough and prevalent enough that "backwardness has been taken up as a key feature of queer culture" (7). By Love's understanding Andersen is certainly queer: a strange, lonely, melancholic outsider.

Andersen's impossible, failed romantic relationships, especially his unrequited passion for Edvard Collin, queers him further. Andersen frequently wrote to Collin and the intensity of his longing and need for Collin to return his affections pervades his letters. In one, Andersen writes, "I long for you; yea, I long for you [...]. I never had a Brother, but were I to have one, I could not love him as I love you and yet – You do not reciprocate! it torments me" (Andersen Letter 88).[2] Four days before Collin's wedding, Andersen wrote to congratulate him on his nuptials. Yet in the same letter the author laments his own loneliness: "I will spend my life alone [...] but give me as much as you can, you are the one I love the most" (Andersen Letter 98). Collin "was naively flummoxed [...] at first" but later more firmly expressed his inability to "satisfy Andersen's wishes" (Bellot 2019). It is Andersen's refusal to relinquish his love for Collin, even in the face of rejection, that makes him backward and, thus, queer: without Collin he is heartbroken, lonely, melancholic, longing for a love utterly out of his reach.

It is little wonder, then, that multiple researchers argue for a biographical reading of "The Little Mermaid," with Andersen as the forlorn mermaid and Collin as her human prince (Bellot 2019; Meyers 2001; Thomsen 2021), and his queer affect permeates the tale. In Jean Hersholt's translation, the little mermaid is repeatedly described as "an unusual child, quiet and wistful" (Hersholt 2019 translated from Andersen 1837). While her sisters are content with their lives under the sea, the little mermaid longs to join the world above the surface. On her fifteenth birthday, the mermaid is finally allowed to see the human world for herself. She rescues a human prince from shipwreck, thus saving his life, and it is then that her longing to be amongst humans transforms into an abiding love for him. Yet this love only deepens her melancholia because the prince "[does] not even know she had saved him" (Hersholt 2019). As her love for humans and her human prince grows, she eventually "[steals] out of her father's palace and, while everything there was song and gladness, she sat

2   Danish to English translations provided by Torsten Thomsen and Charlotte J. Fabricius.

sadly in her own little garden" (2019). Here the little mermaid's queer back-wardness is thrown into sharp relief: while her family is happy and celebrating together, she stands isolated and alone in her grief. It is this grief that drives the mermaid to visit the sorceress, who agrees to transform the mermaid's tail into human legs in exchange for her voice.

It is her prince who discovers her when she washes ashore. Though he brings her to his castle, Andersen's little mermaid still cannot find relief from her lonely melancholy. Rather than a companion or a lover the prince might share his life with, the mermaid is treated like a pet, never fully accepted as his equal or even as human. She remains an outsider in the castle – and, by extension, in the human world – as much as she was under the sea. The little mermaid entertains a fleeting hope that the prince will marry her, yet that hope is quickly dashed when the prince reunites with the princess he believes saved his life. The mermaid suffered his implicit rejection in isolated silence. She cannot share her torment with anyone, least of all the prince, and she must quietly watch as he falls in love with someone else. The little mermaid dances at the prince's wedding and "[e]veryone [cheers] her, for she had never danced so wonderfully" (2019). Yet her "tender feet [feel] as if they were pierced by daggers, but she [does] not feel it. Her heart suffered far greater pain" (2019). Much like in the palace under the sea where she isolated herself from her happy family because of her melancholic longing to live as a human, the little mermaid is once again alone with her suffering. In the end, the little mermaid chooses to sacrifice herself to keep her prince safe rather than kill him so she can return to her family beneath the waves.

"Part of Your World": Ariel's Longing, Ashman's Hope

Of course, Andersen's "Little Mermaid" has been told and retold dozens of times since the original tale was published in 1837. Yet few iterations are as famous as Disney's 1989 film, *The Little Mermaid* (dir. Ron Clements and John Musker). On the surface, Disney's adaptation is another heteronormative story of a young princess falling in love with and eventually marrying her prince. However, a deeper examination reveals *The Little Mermaid*'s queer genealogy, originating with Andersen's own loneliness, longing and melancholia, and continuing into the animation studio's history.

Sean Griffin points out that Disney's animation studio has always had a space for outsiders and those who do not quite fit in, even if they were not explicitly queer. Like LGBTQ+ people, these "characters do not find another world to escape to and must confront their ostracisation. Rebuffed by the upholders of what is normal and decent, these 'rejects' eventually find acceptance and happiness by finding a use for those aspects of themselves that were originally thought to be deficient or 'abnormal' " (Griffin 2000: 64). Openly LGBTQ+ employees "have [also] at times expressed the importance of their sexuality in the work they do, indicating that they have expressed outlooks and opinions about sexuality in their work for Disney" (Griffin 2000: 133–4). This means that even if Disney's animated film canon predominantly upholds heteronormative and cisnormative hegemonies, the queer affects, sensibilities and aesthetics of Disney's LGBTQ+ creative talent are still legible in the animation studio's filmic canon. *The Little Mermaid* is no exception. Andreas Deja, an openly gay animator, and Howard Ashman, a gay man with AIDS who frequently produced and provided lyrics for Disney's Eisner-era films, both lent their talents to *Little Mermaid*. Deja "has announced in various interviews that his sexual orientation has had its effect on the characters he draws," (Griffin 2000: 141) which is particularly apparent in his work as the lead animator for Ariel's father and king of the seas, King Triton. Griffin notes that Deja's animated men, including Triton, "present an exaggerated ideal of the masculine body, with massive shoulders, chest and arms, tight waists and rugged facial features" (2000: 142). Such a queer aesthetic is not limited to Deja's work. Ursula, the sea-witch who grants Ariel a pair of human legs in exchange for her voice, is modelled on the drag queen Divine and Ursula's "campy nature is due at least in part to the words that Ashman gives her to perform," (146) particularly in her solo number, "Poor Unfortunate Souls."

Yet it is Ariel's own queer affect – her loneliness and her longing, passed down from Andersen to his mermaid to Howard Ashman – that is most prevalent in the film. Ashman, a gay man who died in 1991 due to complications with HIV/AIDS, was intimately acquainted with backward feelings like alienation, isolation, loneliness, longing and melancholia – all evident in

his work on *Little Mermaid*. Griffin notes, "Although Ashman was not the
arbiter or final decision-maker on what projects the studio would produce
as animated features, the three projects Ashman was involved with [*Little
Mermaid*, *Beauty and the Beast* (1991), and *Aladdin* (1992)] are all adaptations of popular children's tales that already had a history of importance in gay
culture" (144). Each is focused on the outsider – Ariel, Belle and the Beast,
and Aladdin, respectively – longing for and finally finding a place where their
queerness is embraced and accepted. Ashman's queer affect is legible in all
three films, particularly *Little Mermaid* where "a sad and isolated mermaid
[...] cannot have the human male she loves" (144). Like Andersen's nameless
mermaid before her, Ariel is at odds with the norms of her undersea kingdom.
The film opens on Ariel's voice recital for the entire kingdom of Atlantica,
yet Ariel is nowhere to be found. Instead, she is somewhere in the kingdom's
forbidden, murky depths, exploring a sunken human ship for trinkets and
treasures she can add to her hidden collection. In "Part of Your World", Ariel
longs to be "up where they walk, up where they run/up where they stay all
day in the sun/wanderin' free" (Clements and Musker 1989), voicing "feelings
common amongst homosexuals, emotions of separation and longing for acceptance – particularly acceptance of the love they feel" (Griffin 2000: 151). Like
her predecessor, Ariel saves a human prince, Prince Eric, from drowning and
falls in love with him, even though interacting with humans, much less falling
in love with them, is "the ultimate taboo in her society" (151). However, this
is the point where Ariel diverges from her queer mermaid ancestor. Ashman's
films speak to an optimism that LGBTQ+ people will eventually be accepted
in society and *Little Mermaid* is "structured around resolving [her] unsanc-
tioned love" (151). Rather than dying like Andersen's little mermaid, Ariel lives
to marry her prince. King Triton eventually relents and blesses their union,
fashioning "a symbol of the new ties between [humans and merfolk] by creating a huge rainbow [an iconic symbol of the LGBTQ+ community] over
[their] shipside wedding" (152). Whether or not Ashman's hopeful optimism
for LGBTQ+ acceptance has borne out is still up for debate. Nevertheless, his
and Ariel's queer affect continue to resound today.

   Like so many of his other fairy tales, Hans Christian Andersen's "The
Little Mermaid" is the offspring of his own queer affect, specifically his feelings
as a lonely, melancholic outsider longing for societal acceptance and for the

Queer Affect

fulfilment of his impossible passion for his adoptive brother, Edvard Collin. Andersen's silent, wistful mermaid has given birth to dozens of adaptations and reinterpretations, all of them carrying the seed of her and Andersen's queer affect. Among these adaptations is Disney's 1989 film, *The Little Mermaid*. Drawn to the film due to his own queer feelings of loneliness and longing, producer and lyricist Howard Ashman carried on this queer affect through the lyrics he wrote for the film's protagonist, Ariel, especially in her song, "Part of Your World". Thus, Hans Christian Andersen is the root of a queer family tree: his queer affect generated his little mermaid, who in turn inspired Ashman and Ariel. Tracing such lineages is crucial: while members of the LGBTQ+ community remain backward outsiders, we need to know exactly where to look for our history, community and solace. While Andersen did not identify as queer in his time, his feelings are the same as those experienced by the queer community. Sometimes knowing where to find fellow backward outsiders is enough.

Acknowledgement

My greatest thanks to Torsten Bøgh omsen, Charlotte J. Fabricius, Meg-John Barker, Bee Bentall, Annie Maier and Charlie Ward for supporting this melancholic, backward queer.

Image Intervention II: Untitled

Figure 32. Artwork by Gemma Files (Reproduced with permission).

Part IV

Identities and Difference

Agnieszka Kotwasińska

Into the Drowning Deep
(Mira Grant, 2017)

There are few so thoroughly gendered mythological beings as the mermaids. It is their very femaleness – alluring, if somewhat vague, thrilling but deadly – that seems to define them not only through their name (mer-*maid*) but through their ubiquitous visual representation in Western art as well.[1] While "mermen," "merfolk," and "merpeople" are becoming more recognisable in today's cultural climate, mermaids remain the most iconic examples of these sea creatures: half-women and half-fish, naked or almost naked, with long flowing hair, often singing, and sometimes shown with a comb and a mirror, symbolising both their beauty, vanity and potential deceptiveness. It is the mermaids' gender(ed) duplicity that stands at the centre of Mira Grant's *Into the Drowning Deep* (2017).[2] Monstrous mermaids betray their human victims by *merely* mimicking human femininity in order to hunt them more efficiently, as the mermaids are in fact male, not female, their faux-femininity only a mask and a form of aggressive mimicry. While such a conceptualisation of femininity makes for a great narrative twist, it also forms a part of a larger critical discussion concerning sexual difference, with femininity understood as masquerade (Doane 1991; Riviere [1929] 1991) or mimicry (Irigaray [1977] 1985). Importantly, the novel consistently draws attention to both the mermaids' looks and the

---

1   Historically, interest in mermen predates that of mermaids, with the Mesopotamian fish-god Oannes dating back to 5000 BCE. Greek and Roman mythologies would link mermen (or tritons) with godlike origins, and mermaids with that of beastly creatures (harpies, Scylla, sirens, Nereids, etc.). See also Scribner (2020) and Wood (2014).

2   Mira Grant is Seanan McGuire's pseudonym, under which she writes SF/horror fiction. McGuire is best known for fantasy series *Wayward Children*.

act of looking at them, and in doing so reveals how human desire and pre-occupation with the scopic logic structure the categorisation of beings into humans and non-humans (or monsters).

Early Christian and medieval representations of mermaids married pagan fish-folk images with Christian ideology, which actively sought to dethrone the sacred feminine from the newly christened populations in Europe (Scribner 2020: 104). Thus, these early images established a strong link between femininity and lust, aberration, and deceit, but by doing so they also inadvertently legitimised mermaids and inscribed them in popular imagination (Scribner 2020: 130). According to Frederika Bain amongst a wide variety of medieval human-animal monsters (of any gender), "the specific category of the monstrous as predicated upon hybridity is primarily the realm of the female," with "the lower bodily stratum of women, be they fully human or part animal [...] figured as bestial and infernal" (2017: 17–18). Still, the mermaid's monstrousness is tempered by her melancholic beauty and the polymorphous desire she represents. Human desire for an otherworldly bliss in the arms of an enchanting sea maiden is safely reworked as the mermaid's ambivalent longing for a fully human life; hence, the persistent popularity of mermaid bride stories over the centuries, including the most famous variant of all, the Little Mermaid (Andersen 1837). Not surprisingly, the characters in Grant's novel struggle with both extremes found in popular mermaid representations: ruthless sea monsters luring weak-willed sailors to their doom as well as naïve and harmless almost-girls yearning for human companionship and a way out of the water.[3]

Grant's story, set in 2022, follows a massive expedition on the ship aptly named *Melusine* (a medieval serpent-like female creature) to finally prove the existence of mermaids in the Mariana Trench. A much smaller expedition on the *Atargatis* (a Levantine fish-goddess) ended in tragedy in 2015, with the entire crew lost at sea. Leaked footage found on the drifting *Atargatis* shows the crew killed and eaten by murderous mermaids, but the footage itself has been written off as a failed publicity stunt and a hoax and has since become fodder for conspiracy theorists and cryptozoologists. Both expeditions were funded

---

3   The mermaids in Grant's novel are often referred to as "the lovely ladies of the sea" – this slightly ironic term of endearment reveals the import of their perceived femininity.

by a media company, Image Entertainment, which initially wanted to make a mermaid mockumentary, but is now bent on restoring its good name, proving the *Atargatis* disaster was not a hoax, and of course, making substantial profit in the process. The *Melusine* is a strange animal – a giant cruise ship retro-fitted for this particular journey as a floating lab *and* a hunting reserve. Several hundred marine scientists are joined by Dr Jillian Toth, a famous sirenologist; Tory Stewart, a sonar specialist and a younger sister to one of the *Atargatis* victims; Jacques and Mitchi, world-famous and fame-hungry hunting duo; Olivia Sanderson, a quirky Imagine reporter; the Wilson sisters (an organic chemist, a submersible driver and their ASL interpreter); security teams hired for their camera readiness rather than their skills; the ship's understandably antsy crew, and a group of stressed-out engineers working around the clock to repair a ship-wide shutter system, the last line of defence against the mermaids. Most scientists embark on this journey in the hopes of furthering their own marine research and do not care about the mermaids; only a select few actually believe in them, and of those even smaller number are aware that the creatures in question are dangerous predators, driven out of their deep-water habitats by catastrophic climate changes and lack of food.

Two intertwined themes running throughout the novel are deception and femininity. In her 1929 essay "Womanliness as a Masquerade," Joan Riviere famously put forward that "[w]omanliness [...] could be assumed and worn as a mask" by women who want to hide their masculinity – masculinity which they steal or appropriate from men (94). While Riviere's essay feels painfully dated, especially when it comes to power dynamics underpinning its discussion of white femininity and Black masculinity, Riviere's radical hypothesis that there is no actual difference between "genuine womanliness and the 'masquerade'" (1929: 94) has sparked imaginative reappraisals by feminist critics, chief among them film theorist, Mary Ann Doane. Theorising cinematic representation of women, Doane points out that the adoption of femininity (or a mask of femininity) introduces a distance between the image and the woman, something that is absent in Western conceptualisation of femininity, in which the female *is* the image. As such, "[b]y destabilizing the image, the masquerade confounds [the] masculine structure of the look. It affects a defamiliarization of female iconography" and denies "the production of femininity as closeness, as presence-to-itself, as, precisely, imagistic" (Doane

1991: 26–5). This visual overdetermination of femininity, the sheer scopic excess of it, is semi-ekphrastically explored in *Into the Drowning Deep*, as the mermaids are almost always described through their images: first, in the few shots captured on the *Atargatis* tape, then in the video caught by the underwater cameras. Even when the motley crew of the *Melusine* finally encounter the mermaids in the flesh, many are shocked to discover the sea creatures do not fit familiar cultural representations of mermaids. After all, "mermaids were beautiful, mermaids were peaceful and gentle and kind, mermaids were fairy tale creatures given flesh, and who cared about the Atargatis video?" (Grant 2017: 175). The overdetermined image is thus always there, floating near the surface and obfuscating the truth: the creatures have developed curved ribs suggestive of breasts, a swelling reminiscent of narrow hips, long and bioluminescent hair-like appendages, and "a surprisingly sensual mouth brimming with needled teeth" (Grant 2017: 14) to attract their human victims. However, in a surprising twist, the mermaids attacking the ship in well-organised hunting packs turn out to be males of their species. And yet, even when the *Melusine* crew and passengers are faced with the proof of the creatures' non-female *and* non-human embodiment, the humans are still overwhelmed by the creatures' physical allure: "It wasn't human [...] but it was beautiful. The slope of its forehead, the angle of its cheekbones, the soft, distressingly human pout of its lips [...]" (Grant 2017: 254). Curiously, one can notice a certain semantic slippage at work in this passage: beauty is defined through femininity, which in turn brings the creatures into the realm of human desire, and finally it is the human yearning for the mermaid (or at the very least appreciation of its beauty) that works to feminise *and* humanise the creature. Still, the process is interrupted, because the mermaid is not female enough and not human enough.

Deep in the bowels of the *Melusine*, a group of scientists attempt to bring the mermaids into the human fold on the basis of advanced cognitive and linguistic skills. They work to establish communication with a captured mermaid, hoping that once the creature recognises its affinity with the humans, it will stop thinking of them as prey/food (and conversely, the humans will see it as more than a ferocious animal). In the end, the captured mermaid saves the scientists from its ravenous kin, who leave the humans alone thanks to its intervention. Historically, asking for proof of intelligence, self-awareness, or language in order to assign some degree of humanness to non-human animals or to human Others has always been part of a larger set of procedures, by which the deeply hierarchical categorisation of humanity could be maintained in its rigid form. The captured mermaid thus joins a long list of animals *and* people kept in cages, prodded and gawked at, for the benefit of variously defined social, medical, military, or technological advancements. As Rosi Braidotti eloquently argues, the question of who or what is assigned the status of a human being has been governed by "a criterion of hierarchical distinction that is sexualized, racialized, and naturalized" (Braidotti 2011: 99), which has resulted in a fairly limited definition of "the human." However, in today's late capitalist climate, these very same criteria "act as the forces leading to the elaboration of alternative modes of transversal subjectivity, which extend not only beyond gender and race, but also beyond the human" (Braidotti 2013: 98). This perhaps explains why the novel does not dwell long on cognition, self-awareness, and language as markers of one's humanity but rather focuses on the way mermaids' strange, yet fascinating embodiment destabilises the anthropocentric criteria: they are not easily sexualised, as they are not female, but they are not human males either; they escape the logic of racialisation altogether; and they resist naturalisation as highly intelligent non-human animals (amphibians) with advanced communicative skills.

Apparently, the sirens communicate via three languages: "[s]poken, sign, and stolen" (Grant 2017: 225). Aside from their own sound language, they use a sign language during the hunt, and a mimic language made up from borrowed sounds (human voices, hum of ship engines, whale songs, dolphin clicks, etc.). As Dr Toth, in the novel, explains about the latter:

> It's not echolalia; it's fixing on the words that had the most stress on them, which means it's fixing on the words most likely to be a call for help, or an invitation [...]. The speech is a lure, like an anglerfish's light or an alligator snapping turtle's tongue. It's a lure that specifically attracts intelligent creatures. (Grant 2017: 242)

Grant reworks here a popular association of sirens/mermaids with singing and having a beautiful voice – an association that emphasises the creatures' purely ornamental faux-femininity and deceptiveness. The sirens are then predators that use advanced aggressive mimicry – mimicking humans both through physical resemblance and the sounds they copy. Clearly, aggressive mimicry, understood as a purely biological advantage, pushes the sirens further into the realm of non-human animals, which have developed certain features over thousands of years to increase their chances of survival. At the same time, mimicry stands in close proximity to the very human concept of the masquerade, and as Luce Irigaray argues in *This Sex Which Is Not One* (1985), mimicry has been a historically safe strategy for women to adopt, a way to "convert a form of subordination into an affirmation" (76). Irigaray proposes that a playful repetition of stereotypes associated with women will never be seamless, and therefore it might eventually lead to a potential disruption of a masculinist discourse. Similarly, in Grant's novel, sirens' mimicry – always out of context and out of place – ultimately calls into question the very anthropocentric discourse, by which the humans attempt to domesticate the sirens. In the end, the very categories that humans deploy to comprehend (and control) the mermaids are used against the humans, who continually stumble when faced with the actual creatures.

The Western act of looking posits the male self as the seeing-knowing subject and the female self as the object of his gaze, and for centuries such logic has governed the way mermaids were read as female (and thus imagined as available, alluring, deceptive and dangerous). In Grant's novel, this scopic regime is interrupted not by the masquerade as such but rather by the very reveal of the mask. Once the mask is recognised for what it is – a cunning biological mechanism banking on human obsession with femininity – some of Grant's protagonists insist on switching to "sirens" instead of "mermaids," the assumption being that the siren is less clearly gendered than the mer*maid*.[4] The unmasking takes place during a siren necropsy, when Dr Toth, with Tory's assistance, proves that the sirens are sexually dimorphic amphibians and their vague "femaleness" and "humanness" are both false. After all, their "human female" disguise is just one of many deceptive adjustments in their repertoire.

As Tory, a young sonar specialist, discovers in the final act, the female of the species is a mammoth beast drawn from the ocean depths by the human

---

4   Even though classical siren had avian features, with time it became more piscine, and in a sense the mermaid semantic field swallowed up the siren. It was around the seventh century CE when associations with Eve and the serpent as well as references to Greek and Roman mythologies began to overtake the avian connections. See also, Braham (2014).

flesh offering made by her mates. Ironically, the description of the female is purposefully vague, and readers learn few details about her physical features:

> Below the Melusine, deep and descending deeper, the matriarch swam. She had been close to a healthy feeding when the brightness had come, searing her sensitive eyes, turning her away. She had eaten a full dozen of the males in her anger, and would eat a dozen more before she could be soothed. They knew her anger for the terror that it was. (Grant 2017: 432–3)

While Riviere's "genuine womanliness" is true for the sea creatures, humans are still not able to comprehend it. Or perhaps, the true mer*maid* is a monster too grotesque and horrifying to be assimilated into the Western scopic regime, its real femininity simply too much to handle.

Astrid Crosland

The Little Mermaid (Hans Christian Andersen, 1837)

Of all the aquatic creatures pervading popular culture, there is one who endures in infinite incarnations: the young royal who optimistically exchanges her fundamental nature in pursuit of her desires and discovers her enthusiasm at odds with earthly experience. At once a coming-of-age parable, a tragic romance, a spiritual adventure and a classical fish-out-of-water story, testament to its timelessness, *The Little Mermaid* [*Den lille Havfrau*] (Andersen 1837)[1] has been adapted in numerous inspired retellings. The mutable nature of mermaids makes them an ideal figure to explore matters of identity: they are not quite human but nor are they fully animal, they might do monstrous things, but they are infrequently categorised as monsters, they are dangerous to humans but equally often endangered by them.

Scholars have interrogated the original text and its numerous adaptations focusing of a variety of identities: from bodily integrity and autonomy, as in Yamoto's reading of the "Surgical Humanisation" of the protagonist, detailing the procedural correction of "an unstable compound of earth and sea, nature and religion, body and soul" (2017: 295), to the "poignant parallels between the mermaid's story and the process of identity development among transgender persons" (Spencer 2014), to the construction of heteronormative femininity (Frasl 2018), to international and intercultural retellings of the mermaid myth (Fraser 2014). What remains consistent is that *The Little Mermaid*'s adaptability across context and throughout the centuries is facilitated by the faceted identity

---

1   Page citations for this tale refer to R.P. Keigwins' 1976 translation republished in *The Classic Fairy Tales* (1999).

of the titular character. From her curious birth station to her journey to make bonds beyond the familiar, and her transcendent fate, this chapter shall focus on how Andersen's original text is influenced by contemporary attitudes that illuminate some of the central elements of the protagonist's identity.

By Andersen's time, the European fairy tale had long been established as a literary form, using characters like anthropomorphic animals, nationally unspecific royalty and folkloric creatures as textual elements. Some writers, such as Charles Perrault, further imbued the format with educational value, ending many of his works with a moral that explicitly commented on the benefits of virtue, as well as other less honourable facets of social navigation, as in the case of *Little Red Hood*, where Perrault addresses predatory relationships with thinly veiled commentary: "Young children, especially young girls [...] are in the wrong when they listen to just anybody [...] so many are eaten by the wolf. I say 'the wolf' because not all wolves are of the same type [...] soft-spoken wolves are the most dangerous wolves of all" (Perrault 2002/1697: 137).

While Andersen does not address the reader so directly, his tales are nonetheless moralised, idealising feminine endurance. In an introduction to Andersen's tales, Maria Tatar describes the little mermaid as "the real virtuoso in the art of silent suffering [...] transforming mortal agony into transcendent beauty" (1999: 214) a common trajectory amongst Andersen's protagonists. Additionally, Andersen fairy tales are filled with overt Christian values: protagonists are rewarded for piety and subservience, punished for selfishness and individuality (see *The Red Shoes*, where the wrong colour footwear proves a fatal error). While Christianity is the presumed background of many fairy tales, with references to godparents, Heaven, and the devil, they are rarely as direct in religiosity as Andersen's tales. *The Little Mermaid* is one of Andersen's most subtle narratives, as the conversion of the mermaid to Christian values can be justified by the fact that the story does not begin with her clear understanding of morality or mortality, her immersion in human culture exposing the boons of the human condition, despite the trials that precede any reward.

While it is common for characters to go by a moniker which indicates their function within the narrative as a generic feature of the European fairy tale, mermaids are seldom personalised, instead objectified as either a seductive danger or reward for another (male) protagonist. Indeed, this is implied to be the case in *The Little Mermaid* for all mermaids except the protagonist. It is a

significant point of difference between the little mermaid and her family that
while her elder sisters are initially excited by the surface world, they quickly
tire of it once they are permitted to visit and choose instead to remain in the
deep kingdom, and her grandmother cannot understand why she would want
to leave their world at all. This idea is furthered in the connotation of the original Danish *Havfrue*, which translates more literally to "sea-wife" than "mermaid." While both denote an elemental essence, the latter term, as a cognate of
maiden, implies unattachment whereas "sea-wife" infers a matrimonial obliga-
tion. Just as a bride is expected to take on a portion of her husband's identity
through his name so too does the original Danish ascribe an indelible quality
to the species: mermaids are "of the sea" by categorical designation. However,
the protagonist is the "little" mermaid, her status implied to be conditional
by the diminutive – an inference that she is less of a mermaid than her family,
mermaid-ish, rather than completely so. It is this point of difference that enables the transformation of identity for the character. While maidenhood is
presumed to be a temporary state, marriage is a singularly sacred transformer of
identity within the logic of the text. The sea King is introduced as "a widower
of some years," deferring all feminine domestic duties to his elderly mother
rather than remarry. Further, when the little mermaid first begins to inquire
whether her species could acquire a soul, her grandmother tells her, "only if a
human being loved you so much that [...] he let the priest put his right hand
in yours as a promise to be faithful and true here and in all eternity – then his
soul would flow over into your body" (Andersen 1999: 224).

    This is presented as an impossible scenario, as her fish tail would surely re-
pulse any potential human suitor, a comment that spurs the mermaid towards
a transformative solution. The lack of name is a further clue in the puzzle of
the mermaid's identity. In human society, names are the most fundamental
marker of identity, ascribed at birth to mark one as an individual, as in the
case of a first, given, or Christian names, and as part of a social grouping, as
in the case of last, family, or sir-names.[2] The little mermaid is doubly denied
the individuality a name provides. She has not been christened, living under
the sea without a soul, but is additionally at continual odds with her royal
family and only seldom referred to in connection to her father or as princess.

---

2   'sir-names' is used here to highlight the patriarchal aspect of the narrative.

From the opening passages of the text there is conflict between the mermaid's ascribed being and her sense of self. Before the protagonist is intro-duced, Andersen details the eldritch beauty of the kingdom under the sea, the palace a living composite of coral and cockle, fish and flowers blurring the limits of indoor and outdoor by freely occupying the halls, the scene lit by the sulphur-blue sand and only during "a dead calm you caught a glimpse of the sun" (217). However, to describe the physical features of the main character he chooses references that are resolutely land-based, automatically creating a distance between her nominal designation as "of the sea" and her corporeal form. The physical description of the mermaid gives her skin as "clear and delicate as a rose-leaf" and eyes "blue as the deepest lake" (217).

While some translations favour "sea" instead of "lake" for the Danish *sø*, Yamoto (2017: 299) tracks that within the logic of the text the water is first compared to the cornflower, a delicate land flower, emphasising the persistency of land-based references and the incongruity between the mermaid's "natural" state and her narrative trajectory. Andersen's mermaid is initially defined by her environment, her worldview restricted by undersea customs that prevent her from rising to the surface and the limited information her family choose to share with her.

Her inherent fascination with the surface world further manifests in the one place in her father's kingdom she is permitted to freely express herself. Each princess has an allotment which they are allowed to arrange to their hearts content. While her sisters take inspiration from their submerged surroundings, the little mermaid chooses to model her garden like the sun, an object she has only dimly perceived through leagues of open ocean. In a show of fastidious single-mindedness, the little mermaid permits only scarlet blooms to grow there, with a single ornament claimed from the myriad sunken treasures that find their way to her father's kingdom. In this round, red, valley she orients the gleaming icon of her desires, a marble Prince shaded by a rosy weeping willow. The roots and branches of this tree "play at kissing each other," this interaction mirroring the mermaid's infatuation with the performative acts of love, particularly marriage. While in numerous adaptations the mermaid's primary motivation is framed as her emotional connection with the human Prince, the original text almost always mentions him in conjunction with her overarching goal to possess a human soul. This establishes the Prince as a tool that will facilitate her ultimate desire – the possession of an immortal soul which would someday allow her to rise into the sky, ascending to explore celestial realms no mermaid has ever experienced. As it stands, the little mermaid is horrified by the prospect that when she perishes her remains will become a frothing scum that rises only as far as the water's surface, with not even a grave for her kin to visit, stating: "I would give the whole three hundred years I have to live, to become for one day a human being and then share in that heavenly world" (Andersen 1999: 224). So great is the little mermaid's fascination with the surface world, and the heavenly world beyond, that it drives her to extreme means to achieve the ability to experience it.

Beginning the arduous journey towards this goal, she travels through a monstrous swamp, filled with grotesque polyps who grip the skeleton of a less fortunate mermaid traveller. Successfully reaching the sea-witch, the little mermaid is confronted by her hideous form, aged and "spongey," as if already on the brink of death and collapse into sea foam. These sights only reinforce the mermaid's determination; even the witch's stipulation that each step the mermaid takes on land will bring her intense pain is not enough to deter the little mermaid.

In addition to the profound spiritual lack that vexes the little mermaid, the kingdom under the sea has everyday ordeals too. When the mermaid comes of age her grandmother adorns her in royal attire, eight oysters clipped along her tail fins, pinching hard enough to cause the little mermaid to complain, prompting her grandmother to comment that "one can't have beauty for nothing" (220). Like her sisters, the youngest sea princess is crowned with a wreath of pearls, further enforcing the theme of suffering and beauty being intertwined – as pearls are formed as a response to particles irritating the delicate interior of the mollusc shell. The lily-like design of the wreath further connotes death, and the mermaid notes that the sun-like blooms of her own garden would suit her better. Though this is a suggestion of the mermaid's emerging independence and individuality, she does not yet dare to cast the lily/pearl/wreath aside.

The mermaid's visit to the sea-witch to bargain for legs is a crucial turning point in the story as well as one of the most iconic passages, marking the mermaid's first significant transformation. While "sacrifice" usually connotes a high price or poignant loss, it can also mean "to make sacred" by means of dedication to a deity and exchanging her tongue is an assertion of her "dedi-cation" and the mermaid's desired trajectory towards her ultimate goal of an afterlife in heaven with God; sacrificing her voice, that the witch describes as the "loveliest voice of all down here at the bottom of the sea," (226) sacralises her subsequent journey. The act further resembles a vow of silence, a show of devotion historically undertaken by some devout Christians. Indeed, when she chooses to die instead of committing the sin of murder upon the Prince and his innocent bride, the little mermaid's body is once more transformed, raised to a higher state, becoming ethereal. As she questions her new form, she learns she has become a "daughter of air," one of a cohort given a chance to earn an immortal soul through three-hundred years of good deeds, a term that can be shortened if she happens across good and happy children or lengthened by their misdeeds. This comment is the most direct address to the audience, and while it projects the ultimate fate of the mermaid onto them, it also implies hope for those like the mermaid who seek a status beyond what is ascribed to them at birth, that they too may someday ascend to states beyond their imagination.

As one of the most well-known of Hans Christian Andersen's fairy tales, *The Little Mermaid* has served as inspiration for popular media from jewel-toned, family-friendly, American rom-com (Clements and Musker, *The Little Mermaid*, 1989) to lustrously metallic, explicitly-adult, Polish horror-musical (*The Lure* (Smoczyńska, 2015)). Treatment of the protagonists in each of these adaptations are similarly diametric: The Disney version gives the mermaid the name Ariel, connoting an airy quality and a biblical connection that handily allows the film to circumvent the pronounced themes of identity and spiritu-ality present throughout Andersen's original. *The Lure*, however, turns the singular mermaid protagonist into a sister-pair of mermaids, one of whom embodies the tragic romantic idealist, and the other the vindictive man-eater. *The Lure* retains some of the more tragic and grotesque elements present in the 1837 tale but further embellishes the plot with other European folkloric elements, the serpentine bodies and carnivorous disposition of the mermaids evoking the lamia and *rusalka,* as well as sirens which are frequently conflated with mermaids in modern storytelling. *The Little Mermaid* remains a multi-media cultural touchstone, retaining enduring appeal in an ever-adapting re-lationship to the construction of identity, a concern that permeates historic as well as contemporary society. It has unsurprisingly been dubbed "an icon in mass culture [...] the official image of Denmark" (Mortensen 2008: 437) and the statue of the protagonist in the Copenhagen harbour by Edvard Eriksen is a popular tourist attraction as well as a repeated site of protest, having been defaced numerous times. While the core focus throughout Andersen's text is the protagonist's quest for spiritual achievement, she has become representative of self-determination identity on levels from individual to international.

Alison Sperling

"The Mermaid" (Hanna Cormick, 2018)

Hanna Cormick's performance "The Mermaid" opens with the artist lying flat on her back on the concrete ground, wearing a full-face respirator.[1] She is connected to an oxygen tank and IV saline drip, hung on the back of a wheelchair nearby. Her head is adorned with a crown of white, plaster-cast seashells and starfish, an iridescent mermaid tail extends from her bare waist, and Cormick's measured voice begins: "When no one sees you, you can pretend the difference does not exist; each new revelation to each new person is a new wound, a re-becoming different."[2] Viewers immediately understand Cormick's body as a site of vulnerability to the violence of the able-bodied gaze, which imbricates bodies differently in a complex system of viewership. If in solitude one can fantasise about so-called normality, in the performance space and under the eyes of spectators, "each new revelation to each new person is a new wound," a repeated exposure of difference.

Cormick is a performance artist based in Canberra, Australia, who lives with Ehlers-Danlos Syndrome, Mast Cell Activation Syndrome, Chiari Malformation and Dysautonomia, amongst other complications, which require

1   Hanna Cormick is a performance artist with a background in physical theatre, dance, circus and interdisciplinary art. Her work has been performed in Australia, Europe and Asia. Her current practice is a reclamation of body through radical visibility. To learn more about Hanna Cormick's work: https://www.hannacormick.com/current-works.
2   Text of the performance is written by Hanna Cormick. The performance debuted in Canberra, Australia in 2018. The work is not typically experienced as a video work, but I am grateful for the chance to see what would normally be experienced as a live performance.

the use of a wheelchair, respirator and orthoses. "The Mermaid" was her first work as a disabled artist,[3] and is a confrontation between living with a rare disease and the climate crises (Cormick 2021). The work stages her experience of living with these conditions as a process that she undergoes as the performance is underway, slowly transforming from a narrative of self-blame to one of empowerment through illness, fragility and transformation.[4] This shift occurs through a growing recognition that her body is somehow in sync with the deterioration of the planetary body:

> I wasn't listening to the tremors that were running through my cells / That were the same tremors running through the coral, / the seabed, the roots / That we are not on the earth, but of it. / I didn't understand / That my veins were as polluted as the rivers / My lungs full of plastic and petroleum / Pesticides soaking into my fat like the soil / And switching on dangerous genes. (Cormick 2021)

Pollution, plastic, petroleum, pesticides. These are the substances of the waters of the Anthropocene, contaminated waters that yoke bodies together in toxic times. Cormick suggests that these contaminants, at least in part, tie us to the world: "[W]e are not on the earth, but of it." Perhaps she is even more radically distancing herself from the category of the human altogether, demonstrating the ways in which living with disability is often already to be outside of the bounds of the oppressive category of human-ness. Like the coral, like the rivers, like "the rocks peeking from the mountainside," illness unites her with the non-human. The hybridity of the mermaid figure is also temporal paradox: a speculative, crip-futurity (mutate, transform, adapt) as much as a harkening back to an evolutionary, Holocene, pre-industrial past to which she imagines returning (we were all fish once). Her hybridity is thus complexly layered: human, fish, organic, contaminant, technological.

An interlude in the video alerts us to the precarity of Cormick's body out in the open – the first time she performs this piece, she has a seizure. Standing behind her and holding large cards with text, her collaborator Christopher Samuel Carroll blasts a track by *The Everly Ills* as he flips though information for spectators while Cormick seizes on the concrete ground. The cards

---

3   In a 2018 interview for the website *Stance on Dance*, Cormick says of "The Mermaid" that "[t]his work was my coming out as being disabled." See Wiederholt (2018).

4   Cormick is here influenced by thinkers such as Audre Lorde and Paul B. Preciado.

read: "The artist is having a seizure," and "[i]t is an atypical allergic reaction caused by food in the air [...] [o]r food hidden in your bag. Or your makeup. Or your hair product." And as the seizure subsides: "Air is a shared resource but it is too polluted with contaminants for her to breathe." Herein lies the tension of a contaminated world that can both link different beings together while also dividing them or requiring mediation between them. Carroll walks over and pushes the wheelchair to where Cormick lies. She sits up, pushes herself up from the ground and into the wheelchair, and continues the performance: "There's no barrier between us and the world [...]" (Cormick 2021).

Fascination with mermaids and other hybrid water creatures derives from centuries of folklore and myth across cultures and geographies, and maintains a firm footing in contemporary literature, film, television and other media. In their introduction to *The Penguin Book of Mermaids*, Cristina Bacchilega and Marie Alohalani Brown (Bacchilega and Brown 2019: xii–xiv) write that continued fascination with mermaids comes from something "deeply unsettling about a being whose form merges the human with the nonhuman," they are "alluring, but can also be frightening." Their hybrid form is described as a "sign [...] that often serve[s] both as admonition for humans not to cross borders and as incitement to do so." They go on:

> But what does our fascination with this dangerous yet desirable other suggest about us? [Tales of mermaids] reflect our fascination with and fear of female bodies and of water and our dread of predators or poisonous creatures that live in or near water [...] they encapsulate our beliefs about what it means to be human. [They] admonish humans for testing their place in the social and natural world with nonhuman or monstrous water beings. (Bacchilega and Brown 2019: xiv)

Although their introduction does not address the relation between disability and ecologies questions directly, several of their claims help to untangle this relation. For example, their suggestion that mermaids reflect fear of and fascination with women's bodies, and that mermaid narratives challenge what it means to be human, become even more complex when thought in relation to disability's gendered and environmental relations. The mermaid is at least partly a fish out of water, she rarely manoeuvres seamlessly between aquatic and terrestrial worlds. The move between human and non-human worlds can be thought as a kind of navigation of different forms of ability in relation to the body; what might be considered advantageous in the aquatic world can become a challenge, or mark one as different on land and vice versa. In the context of climate change and the polluted waters that Cormick evokes, the hybrid, not entirely fish nor human, offers neither safety or desirability. If the mythic oceanic once offered reprieve from the troubles on land, the Anthropocene has rendered the oceans, filling with plastics and chemicals and threatened by quickly rising temperatures, as much more troubled waters. We see the mermaid myth as evolving differently in relation to the threats of the Anthropocene just as what we mean by disability and for whom also must be in conversation with a changing and increasingly toxic world.

The mermaid fantasy of human/non-human hybridity also fosters fantasies of accessing the depths of a world otherwise unwelcoming and unavailable to humans. Blue humanities scholars often remind us that the ocean is as foreign a place to humans as the far reaches of outer space (see Helmreich 2009). The oceans remain mostly unexplored, and the deep-sea remains especially inaccessible to humans only able to be encountered with the mediation of advanced and expensive technologies. While Bacchilega and Alohalani Brown claim that the continued attraction to retelling merfolk stories "lies in the mystery [of their] existence and the lasting question, 'Are mermaids and other water beings real?' " (Bacchillega and Brown 2019: xvi), I would suggest additionally that the attraction to mermaids in Cormick's work demonstrates the mermaid's ties to ecologies. In other words, mermaids gain access to the most unexplored and yet most endangered realms of the planet; they are the fantasy figure of ecological access to unseen and uncharted terrain and to new forms of life; or, more dangerously for them, access to new depths for their discoverer to plunder, abuse, or colonise.

While mermaid figures have arguably always provoked the boundaries of dis/ability, the explicit convergence of fish-amphibian-embodiment with experiences of disability can be found in a number of recent works worth pointing to briefly here. *The Shape of Water* (del Toro, 2017), winner of the Venice Biennale Golden Lion award for best film and an Oscar for best picture, tells the story of a mute woman Elisa Esposito (Sally Hawkins) who falls in love with an amphibian-like creature (Doug Jones) whom she saves from euthanasia at a secret government laboratory. While living in her bathtub at home, the creature is discovered to have healing powers which he eventually uses when both Elisa and himself are shot and fall into the water, touching her neck where gills begin to grow as they embrace. In the work of video artist Emilija Škarnulytė, the artist wearing a mermaid suit swims through abandoned Cold War nuclear submarine canals, as aerial footage follows her out further and further to sea. Škarnulytė's work often evokes the figure of a future archaeologist, who she imagines in an encounter with the toxic legacy of nuclear and technological ruins of humankind. Her monograph, *Sirenomelia* (2021) is also the name of a rare and fatal congenital disorder occurring in newborns, where the baby is born with their legs fused together as if a mermaid tail, a condition often attributed to environmental factors. In the short story by Caitlín R. Kiernan, "The Mermaid of the Concrete Ocean," (2015) an aged woman in a wheelchair is visited by a young journalist. The woman was the muse of a deceased artist who painted numerous coveted mermaid paintings, one of which hangs in the woman's apartment. The story openly provokes the associations made between disability and mermaid figures, the woman telling the journalist "[d]on't even bring it up," referring to the childhood injury that left her in a wheelchair. She "has always loathed writers and critics who try to draw a parallel between the mermaids and her paralysis" (Kiernan 2015: 258). But as the younger journalist notes, "[o]nly now, *she's* brought it up" (Kiernan 2015: 258, emphasis in original). Disability and mermaid narratives are provocatively related, and yet as Kiernan's narrator complains, it can also be a tired trope, more importantly an offensive methodology for critics to read actual disabled bodies in the world as myth or folklore.

Savina Petkova describes what she terms "Mermaid Cinema" as that wherein

> the female – and disabled – body queers one of the longest-standing fetishes of the male gaze. It is precisely the hybrid nature of the marine feminine which attracts and repulses, its fluidity intimately tied to female maturation, in the process of which the flesh and bone remain both human and foreign at the same time. (Petkova 2021)

Petkova provocatively signals a queering that is repulsive as well as foreign (or non-human). There are many ways in which the mermaid figure can both disrupt categories of desire as Petkova suggests, and yet she just as often is depicted precisely as an adherence to the strictest categories of heteronormative, masculinist and especially colonialist desire. Recent works like Monique Roffey's novel *The Mermaid of the Black Conch* (2020), sculptures by artist Wangechi Mutu, or Bola Ogun's short film *The Water Phoenix* (2017) are each examples of contemporary works that imagine mermaid figures in direct conversation with colonialism and capital exploitation of Black and Indigenous women. Works by these artists and writers reimagine mermaid figures (at least in the majority of contemporary Anglophone cultural examples) against historically white, feminised and sexualised figures according and in more explicit refusals of stifling cisgender and heteronormative structures of gender expression and sexual desire. The mermaid persists in culture largely because she is so entangled together with racialised, gendered, sexualised and able-bodied conventions at once.

"The Mermaid" ends with Cormick reciting a series of self-declarations. These are announcements of her solidarity with disappearing lands and acidifying oceans. They are her banding together with other marginalised peoples and beings, those who are most effected by the effects of climate change:

> I am the low lying islands we are drowning / I am the sick air, sick ocean, contaminated water, earth / I am the damage we have done to the earth / I am all the people you hide away and pretend do not exist / I am everyone you tread on to stand where you are / I am the canary in the coalmine / I am your own precarity and mortality / I am the carrion and the ink black crows that feast upon it / I am not a fighter / I am the battleground. (Cormick 2021)

In addition to other solidarities, the performance's conclusion is also an alliance with toxicity itself. She is not just "the sick air" – she is "the damage," she is "the carrion and the black crows that feast upon it, she is the battleground." One way to understand Cormick's performance would be to see it as her coming to terms with her body, and finding strength in her connection with the earth, the air and the waters. But my feeling is, that in addition to her own transformation that she recounts in the fifteen minutes of "The Mermaid," she also proposes a shift in how we understand disability broadly speaking, and its possible relation to aquatic and other crises in the Anthropocene. Hydrofeminism has pointed to this fact that we are all bodies of water (Neimanis 2017); we cannot think of the human as separate from the waters of the oceans, rivers, lakes, or of atmospheres. This is a way of thinking in terms of connectedness to the non-human that is important to Cormick (I think), as well as in the environmental humanities and in what has been called elsewhere Anthropocene feminism (Grusin 2017). But in a moment when ecological thinking is drenched with rhetoric of entanglement and the enmeshed, it is important that Cormick suggests that what unites us with the world can also be simultaneously the thing that separates us, and it is not always so simply for better, nor for the worse.

Alison Patterson

Song of the Sea (Tomm Moore, 2014)

With *Song of the Sea* (2014) director Tomm Moore and animation studio Cartoon Saloon continued a project begun in their first feature film *Secret of Kells* (2009): interpreting Irish history for contemporary audiences. *Kells*, set in the ninth century, depicts attempts to preserve pre-Christian and early Christian culture from invading Norsemen. *Song of the Sea* is set in the not-too-distant past of the director's 1980's childhood and brings together grief, loss and disability with cultural history and mythology. Its style evokes traditional hand-drawn animation processes and abstract art as much as ancient Pictish designs.[1] In the film's formal heterogeneity, *Song of the Sea* creates overlapping and simultaneous layers of past and present and this world and fairy world. Based on the local myth of the selkie, the film is legible to transnational audiences as a sea creature fairy tale, and to youth audiences via focalisation through young protagonist Ben, the selkie child Saoirse's brother.

The film depicts Ben's (and the family's) attempts to mourn the loss of the mother six years prior at Saoirse's birth and to navigate Saoirse's mutism and her danger to herself. The film's immediate conflict arises when Saoirse discovers a hidden seal coat and, donning the robe, takes on her seal form, returning temporarily to the sea. When Granny sees the child in human form asleep on the beach, she declares the children better off with her, setting off

1   In an *American Cinematheque* conversation with Tomm Moore and Ross Stewart, Guillermo del Toro compliments Cartoon Saloon on acting as a "bastion" against the 3D aesthetics dominating most animation studios. Del Toro observes that the studio has integrated the ideals of historic animation with new digital tools to produce "what people now deem – horribly – 2D." Moore responds: "I don't really like this word 2D because it's *not* 2D – it's hand drawn, a hand drawn aesthetic and it's an illustrated look but it's not 2D. It's kind of playing with all the D's that you possibly can, you know?"

the children's adventure across Ireland to resolve Saoirse's identities and restore the family.

Neither the selkie tale nor the impetus for this version were easy fits for a children's film. In discussing its origins Moore describes a stunning sight on a vacation from Kilkenny to the west coast of Ireland. Seeing seal corpses where frustrated fishermen had slain them, Moore was reminded of the formerly powerful mythology that "seals could contain the souls of people lost at sea or could even be Selkies; people who can transform from seals into humans [sic]" (Moore 2016: 19). Moore was struck by the loss of connection between the human and animal world that had previously prevented such acts and the loss of connection between contemporary life and Ireland's cultural past.[2]

Anna Katrina Gutierrez argues that the female "Selkie Tale" falls within the global tale type of the "Otherworldly Maiden," a schema wherein a man catches sight of a maiden in human form, desires her and, when she's next in human form, takes away the object that powers her transition to trap her in human form (Gutierrez 2012: 26). The female selkie lives freely as a seal, but she is bound to live in her human form on land when a man steals and hides her seal coat. Only by retrieving her seal coat (and thus her agency) can she return to the sea.[3] In *Song of the Sea* however, the mother's disappearance is a prologue to the children's adventure and the crisis hinges on the traditionally least empowered figure, the daughter.

Through an Irish Studies lens, *Song of the Sea* might be read in terms of:

- Sympathy for the heterogeneous figure in Irish folklore and political imagery (see Och 2017).

2   Moore reflects the mood of Seamus Heaney's introduction to David Thomson's *People of the Sea*. Heaney describes Thomson's depiction of the closeness of human and animal via what Edwin Muir would describe as "that long lost, archaic companionship ..." (xiii).

3   David Thomson recounts the sharing of a Faroe Islands proverb – "She could no more hold herself back than the seal wife could when she found her skin" (200). See also *The Secret of Roan Inish* (Sayles, 1994) for a film version.

- Sympathy for the feminine (Ireland and Irish sovereignty have been depicted as female,[4] though female sovereignty even within origin myths is circumscribed).

Song of the Sea might be read simultaneously through a Disability Media Studies lens. The historic literary relationship between disability and monstrosity and the *Song of the Sea*'s selkie as sea creature and mute child invites such a reading. Indeed, Saoirse's mutism and her status as a water creature are intertwined. The film explores Irish identity and familial bonds through Saoirse's mutism as well as her duality as a selkie. Further, a disability studies lens encourages viewers to consider the "unruliness" of her body and behaviour and the heterogeneity of the text as challenges to a hegemonic "normal," (see Davis 1995) though a normalisation of both in the end should be interrogated (see also Mitchell and Snyder 2001).

    Song of the Sea complicates its diegesis at the outset, as the film opens with a dreamy credit sequence. In voice-over, a woman invites a "human child" to "come away" to the fairy world, and we see images that appear to be illustrations but are paintings made by the mother Bronach and her child Ben on Ben's bedroom walls. Vignetting blurs the edges of the screen, and the image dissolves to the exterior, where a bob of seals watches the house high on a rock island at twilight. Inside the house, Bronach guides Ben's painting as she teaches him a song in Irish. Ben tells his father that their selkie painting must be completed before the baby's birth. Throughout this scene, the camera refocuses and the frame edges blur and sharpen; the result is that the figures seem in and out of the present and past, story world and the "real" world of the diegesis which has not yet been settled.

    Bronach plays a tune briefly on a nautilus shell before she gives it to Ben for safekeeping. Concentric circles enclose and embrace the small family as we watch from a birds' eye view. The floor of the room and their home, the ground of the island and the sky, and finally the sea, rotate around them. Ben is tucked in, and we see Bronach from his point of view as she reassures him that he will be "the best big brother." His lids close and the screen goes dark – we are in his optical point of view – and when his eyelids open again his mother is in labour.

---

4   Ireland is still called Erin/Erenn/Eriu for one of the three goddesses of the island.

The credit sequence ends under the sea. The camera tracks upwards to reveal an older Ben and his dog Cú on the beach with 6-year-old Saoirse, "someone who doesn't talk" but is "really loud." Saoirse walks towards the water, approaching gathering seals. In Ben's effort to prevent her from going into the sea, he is pulled under shallow waves. In that scene on the beach, we see Ben has been parentified. This is in one respect, and one only, *a* reality and the "now" of the film.

On the evening of Saoirse's birthday, she follows *soilse* (fairy lights), an important narrative and graphic feature of the film – bright points on the watercolour textured images. After compositing, this design choice makes the lights more material than the figures and the "human" world they occupy. Saoirse discovers a key, retrieves her hidden seal coat and transforms into a small white seal without giving any indication that she knows that her mother before her had been a selkie. In a beautiful underwater sequence, seal-Saoirse, though illuminated, is a tiny figure among other seals, and fish, abstract geometric jellyfish as well as a more representational whale nearly the length of the full screen's width. In between these formal poles, seal-Saoirse is as "real" as the other seals but merely brighter, and her perspective is never privileged over our total view of the ocean environment. Here her mutism presents no problem, for Saoirse or viewers, as language and narrative drive are both suspended with all the film's humans absent.

The underwater reverie ends, and Saoirse washes ashore asleep to be discovered by her panicked Granny. Granny reacts by taking the children from their home, their dog and the sea in a sequence illustrated by dissolves between images of the journey from the vantage of the car and Ben's marking of it on a map he will later use to try to take them home. Before long, Granny bypasses Samhain revellers and installs the children in front of an old radio in her city home stuffed with middlebrow curios and antimacassars.

At Granny's, Saoirse's actions continue to speak where she cannot or will not: she soaks the bathroom and her grandmother's best coat to try to reproduce the conditions of transformation into her seal form in the bathtub. Granny responds by sending the children to bed during daylight. Ben laments facing mounting consequences of his sister's needs and actions. Ben's desire to escape their grandmother's clutches is even stronger than Saoirse's: they may be grounded, but Ben is determined to return home. Saoirse is close behind, the musical nautilus shell from their mother in hand.

This is also the night when the veil between the spirit and human worlds is thinnest, and fairy lights are all around. When three costumed *daoine sidhe* (fairies) accost the pair to guide Saoirse underground, Ben is dragged along, this time literally as he had attached Saoirse to himself with a makeshift leash. The fairies tell Saoirse and Ben that they need Saoirse to become a selkie to call all the fairies "home," and separate the fairy and earthly realms. Ben tells the fairies that she can't sing because she can't talk, but we understand before Ben, who speaks for her, that her mutism is tied to her selkie coat. Not only is she stranded on land without the coat, but also, she is mute because she is Other and her Otherness has been denied. There is a further implication: if Saoirse is to undo the entanglement of "this" world and the fairy world, then Saoirse's very being is a problem to be solved. Here the narrative trouble of earthly and fairy entanglement contradicts the visual theme of simultaneity.

Saoirse wants to go home, though she is also compelled to recover her selkie coat and return to her seal form. In this she is unlike Disney's *Little Mermaid's* Ariel, driven by the romantic ideals of the Hans Christian Andersen source text but spared the mutilation necessary for Andersen's mermaid to transform and the tragic ending of Andersen's tale.[5] Saoirse follows the lights and tries to liberate fairies from their rock forms, despite Ben's certainty that he and his map know better than Saoirse. In the meantime, the selkie is endangered by her separation from the coat: without the seal coat, Saoirse weakens, and Ben's desire to return home for his sake is overtaken by his desire to protect his sister. Nevertheless, it is the younger, mute female child who knows to use leaves to take the sting out of nettles, and who knows that a Holy Well is a portal to a watery pre-Christian Otherworld. Saoirse's special capacity to understand the fairy world, which Ben has struggled to comprehend, is complicated – her otherworldliness is explanatory, but it is near to the familiar "supercrip" trope (see Schalk 2016). Mute, she is supernaturally attuned to the fairy lights that draw her towards the coat and home.

5   Disney's *Little Mermaid* is commonly read as a conservative film, an "advertisement for traditional marriage and family values." See, for example, Deborah Ross' contrast between Ariel and Miyazaki's Ponyo in "Miyazaki's Little Mermaid: A Goldfish Out of Water."

Ben encounters the Great *Seanachaí* (traditional storyteller) in an underground cavern and learns about the connection between Saoirse and the lights, and the threads of myth and narrative that preserve memories through time, literalised as the storyteller's hairs. As he clings to his own narrative thread, Ben sees a projection on the wall of the cavern, revealing his mother's sacrificial return to the sea and Saoirse's birth. In an unmistakably reflexive moment, we see Ben behold his and Saoirse's lives. In the next act, Ben will have to see his sister's capabilities differently.

Now, the children's attempts to return home and recover Saoirse's coat are also compromised by the witch Macha, drawn from Irish mythology but transformed by Moore from raven-like to owl-like. Macha is hybrid, a thematic "shadow Granny" who literally jars feelings, and a visual palimpsest, a Granny-owl-witch all at once. The children free Macha/Granny from her own trap before she provides the aid they need for their final stretch. Early in the film Saoirse is defined by her "lollipop" form and "cute" features, and here that form is overlaid with a deteriorated one as she greys and weakens and takes on a dual appearance of young human Saoirse and an elderly dying Saoirse. That duality is distinct from the ontological heterogeneity of the selkie.

Recall that selkies are unlike mermaids who have features of humans and aquatic beings at the same time.[6] The selkie is spiritually dual but can only exist *physically* in one form or the other. Saoirse is the least physically hybrid of fantastical beings. When the story of the god of the underworld Mac Lir ("Son of the Sea") is reworked within the film, we see Mac Lir having been turned into a stone mountain to "save" him from his grief. He is represented as man and mountain at once: not metamorphosis (which is temporal, a transition from state to state) but portmanteau (which is spatial, carrying the meaning of two things).[7] This is similar to the fairies who are in some frames both living (animated in the literal sense) and stone, but in their case they were *becoming stone* within the present narrative, while Mac Lir has lived in the mountain in mythic time and will be set free within this diegesis.

---

6   Selkies may be inhumanly beautiful in their human form, though physical deformity is attributed to descent from selkie and human parentage. See examples in *People of the Sea*.

7   Here I refer to considerations of animation and metamorphosis in Sergei Eisenstein's *On Disney*.

When Ben restores Saoirse's coat to her over their father's objections, Saoirse rises, spins and transforms from ill and pale to healthy and vibrant in a channel of light before a cut to the bob of seals watching. This is followed by images of fairies released from earthly forms across the landscape. We see Saoirse again from a bird's eye view, her face in close-up facing upward as she sings the titular "song of the sea" to release the fairy spirits. In this ecstatic moment, Mac Lir is released from his grief and his mountain form, and his spirit joins other fairy beings, but the mountain remains. In the release of his grief, he is thematically aligned with Conor, the father. Similarly, the witch Macha, is released from her concrete form and able to transcend.

Only Saoirse is (within the diegetic world) of two alternating forms by nature. She chooses to speak, and she chooses to be human – in contrast, Cartoon Saloon heroines *The Secret of Kells* (Moore and Twomey, 2009), Aisling[8] and Mebh and Robyn of *Wolfwalkers* (Moore and Stewart, 2021), remain fluid figures. This is especially worth considering given animation's graphic shape-shifting potential, where hybridity is not only possible but also a medium specific capacity. In Disney's *Little Mermaid* (Clemente and Musker, 1989) the imperatives of realism win out over animation's potential for fluidity, sparing Ariel not only from pain but also from monstrosity. By contrast, Hayao Miyazaki's eponymous Ponyo's body (2008) evolves and de-volves, a "plasmatic" being in Sergei Eisenstein's sense. *Song of the Sea* asserts that her very being is both the film's distinct pleasure and its problem.

At the end of *Song of the Sea*, Saoirse, a child – daughter and sister rather than captive bride (the usual fate of selkie women) – chooses land and family. It is less that Saoirse has "overcome" her mutism than that she has voluntarily traded her difference to unify and heal her family. This presents a problem: to disentangle the magical and mythic from the "real" is to agree that these categories should be distinct. Saoirse embodies the memory of Bronach and Ireland's spiritual dualism without the challenges presented to others by non-normate being, the film suggests. However, the necessity of the disentanglement and performance thereof also finally affirms the reality of experiences outside normative experiences that a locked or sunken coat had wished away.

---

8   She is drawn from *aisling* poetic tradition in which the poet dreams of a fairy-woman who is both fairy *and* woman, and traditionally read as an allegory for Ireland.

Jon Hackett

Possession (1981) to *My Octopus Teacher* (2020)

In the recent Netflix-distributed documentary, *My Octopus Teacher* (Pippa Ehrlich and James Reed, 2020), the South African filmmaker Craig Foster muses early on: "A lot of people say an octopus is like an alien." However extra-terrestrial the cephalopod may appear, familiarity breaks down barriers between Foster and the octopus of the film's title, as it proceeds. The documentary records the unlikely friendship between human and mollusc, as Foster gradually wins the confidence of the initially shy animal. The emotional charge evident in Foster's attachment led to amused speculation in the press and on social media as to whether this strayed over into something libidinal. Elle Hunt (2020) reassured readers of *The Guardian* that the relationship was no cause for concern. The documentary itself claims that the unlikely attachment served as a sort of therapy for Foster after a period of emotional turmoil.

My *Octopus Teacher* has won various prizes, culminating in the Academy Award for Best Documentary. It was recognised by the British Academy and the Guangzhou International Documentary Film Festival too. In this respect, the film might be read as one of a select canon of transnational films that engage with cephalopods, which also raise the spectre of cross-species attraction. *My Octopus Teacher* is a documentary, where the other films I will consider here are fiction features. Nonetheless, the allegedly alien nature of the beast and its irresistible appeal for humans, as well as the distinctly transnational nature of the film in terms of production, distribution (including festival success) and consumption, are very much features of this and the other films considered. The border and boundary-crossing nature of the cephalopod further seems to serve in the films below as a figure for media that traverse national, zoological and generic boundaries, operating between auteurist "high culture" and "disreputable" exploitation genres of cinema.

The Ur-film in this respect is Andrzej Żuławski's (1981) *Possession*. Shot after the director left Poland, this is the work of a self-imposed exile, who would go on to make most of the rest of his films in France. However, as Michael Goddard highlights, *Possession* is a truly transnational film shot in a sort of transitional state: "[I]t is a film conceived of in Poland; shot in Berlin with a French producer; starring actors from France, New Zealand and Germany; and based on a script by an American and funded by American money" (Goddard 2014: 245). Goddard alludes to the casting of Isabelle Adjani and Sam Neill in the lead roles in this film in which they play a couple whose marriage is under strain. Its transnational reception is also striking, with Adjani winning best actress award at the 34th Cannes Film Festival; the film having a sort of grindhouse release in the United States; and finding its place on the video nasty list in the UK. The film almost defies description in its mixture of heightened expressionist acting, elements of body horror and allusions to Cold War contexts. One can well see that the actors were reputedly near to breakdown from the extreme agitation of the performances, which at times depict violence and self-harm; Neill later claimed: "It nearly killed me, but work with him was great" (Neill 2016). As Michael Atkinson memorably argues of this director *maudit*, "if an argument can be made for him, it would necessarily be in the form of a bludgeoning harangue" (Atkinson 2003: 39).

For our purposes here, note the foundational plot of Isabelle Adjani's character, Anna, who takes up with a shape-shifting creature that alternately resembles a murderous phallic squid, a tentacled alien lover and later Marc's (played by Sam Neill) doppelganger. Anna flees their flat in the Berlin tower block[1] to a run-down apartment in which the creature dwells. She feeds it on people who come into the flat to investigate, including the private eye that Marc has hired. Towards the end of the film, we have a Cronenberg-like image of Anna in the embrace of her tentacled lover, which Marc stumbles upon in an almost delirious state. At the very end, the creature has apparently become Marc's uncanny double. Anna has previously stated her inability to renounce

---

1   Goddard (2014: 245) cites Żuławski's statement that it was the nearest the exiled Polish director could find to Soviet bloc architecture.

this relationship, which is evidently at the level of an irresistible compulsion. But the film apparently entertains the possibility that her lover is merely a fantasy of Marc's, an interpretation that would account for its mutability throughout the diegesis.

Possession, then, might be the template for a restricted number of later films that explore cephalopod–human relations, whether centrally or marginally. The most "literal" successor would perhaps be *The Untamed* [La región salvaje] (2016) by Amat Escalante, which replicates the plot of a tentacled beast that proves to be compulsively addictive for the characters in the film. This Mexican film is transnational both in its elaborate co-production and financing arrangements (twenty production companies and funding sources are listed in the opening credits), and in its intertextual adoption of the narrative and visual tropes that recall *Possession*. Like the earlier film it met with success at festivals, with Escalante winning the Silver Lion for direction at the 73rd Venice Film Festival.

Escalante's film begins with Veronica (Simone Bucio), a young woman who has regular liaisons with a mysterious creature in the woods. The very first shot, however, is apparently of an asteroid, implying an extra-terrestrial origin for the cephalopod. Throughout the rest of the film, Veronica introduces a young mother, Alejandra (Ruth Ramos), her brother Fabian (Eden Villavicencio) and her husband Ángel (Jesús Meza) to the beast. Fabian and Ángel are themselves conducting a secret relationship behind the back of Alejandra. However, Ángel is arrested after the body of Fabian is found lifeless in the woods, a victim of the beast. Veronica next introduces Alejandra to the tentacled creature as a sort of consolation for her marital strife; with the former a replacement for Veronica, whose encounters with the monster have become violent and debilitating. When Ángel is released from prison after initial suspicion of guilt for the death of Fabian, after an altercation with Alejandra in which he accidentally shoots himself in the leg, Alejandra drives him to the countryside for a liaison with the beast. On a couple of occasions, we see similar scenes as in *Possession* of the coupling between human and beast; we also see the beast coiled in the rafters waiting to descend upon the hapless Ángel, the latter with an ecstatic expression on his face.

The cephalopod beast in these films clearly figures exorbitant and excessive enjoyment, as well as compulsion or addiction – the protagonists find the attractions of their "alien" partners irresistible. We might agree with Laura Antón Sanchéz (2020: 73) that in the cinematic encounter between the woman and the monster, "female identity is defined by means of a prohibited desire – empathy with the Other, the one who is different." Previously I have argued that "the monster is the ultimate *polysemic* text" (Duffett and Hackett 2021: 12); clearly here the beast is calling out for interpretative frameworks to account for the nature of the gratification it promises. One such theoretical approach would be psychoanalysis, specifically Jacques Lacan's discussion of feminine *jouissance* (enjoyment), in his 1972–3 seminar, *Encore*. This is especially the case given that female characters, Anna in *Possession*, as well as Veronica and Alejandra in *The Untamed*, are linked (and spectacularly displayed) with the beast. In the latter film, Fabian and Ángel have much briefer and fatal encounters with it, implying that there is something fatal about the creature that overwhelms the fascinated male.

Lacan himself alludes to Freud's famous bafflement before the question of female desire, *Was will das Weib?*[2] (Lacan 1998: 75). In this year of his seminar, Lacan is elaborating on his teaching, essaying the notion of a specifically feminine *jouissance* that escapes the masculine "phallic" gratification theorised by psychoanalysis. Specifically, for Lacan, the woman has "a supplementary jouissance compared to what the phallic function designates" (Lacan 1998: 69). It is this avoidance of characterising female gratification as merely completing a binary, but something that escapes and avoids the zero-sum game of male phallic enjoyment, that has Lacan dub feminine *jouissance* as *pas-toute*[3] (Lacan 1998: 36). In relation to *Possession*, such a conception allows us to account for Marc's (following Freud's, perhaps) bafflement towards Anna's desire. It also provides an interpretation of Anna's enigmatic words when Marc catches her in the rapturous embrace of the beast: "Almost ... Almost ... Almost ...," which we might gloss here as "not-whole" or *pas-toute*. Whether we read this as insightful, uncomprehending or just plain sexist on the part of the male director, screenwriter (or analyst), for Lacan, the radical incompatibility of feminine with phallic *jouissance* will mean that "there's no such thing as a sexual relationship," a repeated refrain throughout *Encore* (Lacan 1998: 12).

2   "What does woman want?"
3   "Not-whole."

The allegedly radical gulf constituted via sexual difference is figured in this film by an alien, tentacled beast.

In the case of *The Untamed*, we might make a similar case to account for the compulsion on the part of Veronica and Alejandra, contrasted with uncomprehending annihilation on the part of Fabian and Ángel, along these gendered terms. The beast is a swarming mollusc that figures "alien" feminine *jouissance* to the male characters and perhaps notionally male spectator. However, in this film that includes the attraction between the brother and husband of Alejandra (not to mention the inherently queer attraction for the cephalopod in these films generally), we might look for less heteronormative schemes to interpret the film (even if Lacan (1998: 71) clearly states that men can experience feminine *jouissance* too). The setting of *The Untamed* in the woods and by the river, allows us to locate it in what Cameron Clark demarcates as the queer anti-pastoral. This will allow us to interpret Escalante's film according to the "more non-egalitarian, inhospitable, and discomforting representations of queerness within the natural world that often struggle to achieve interpersonal or ecological connections" (Clark 2019: 212). The alien, perhaps alienating nature of the spectacle, along with the sterile expenditure that increasingly debilitates Veronica, might point to this reading strategy as (un)productive here.

If these two films in particular appear central to this restricted canon of films, we can point to other auteur-driven transnational cinema that include a more tangential relation to cephalopods in the diegesis. I will mention three films from South Korea here in this regard. First, a couple of films by Park Chan-wook. In *Oldboy* (2003), one of the central relationships in the film is that between Oh Dae-su (Choi Min-sik) and his daughter Mi-do (Kang Hye-jeong). Imprisoned for fifteen years, on his release Dae-su fails to recognise Mi-do as his child and due to a situation deviously engineered by the villain, Lee Woo-jin (Yoo Ji-tae), he enters into a sexual relationship with her. I will note here that the initial encounter between the two after his release occurs in a sushi restaurant where Mi-do prepares him a meal. Famously Dae-su consumes a live octopus whole at the counter, upon which he promptly loses consciousness. This excessive spectacle of oral *jouissance* sets up the subsequent incestuous relationship that is to follow. Although not a chief actor in this film, the cephalopod once more provides the spectacle of obscene enjoyment that is to drive the narrative henceforward. It is also one example of how, for Kyung Hyun Kim, *Oldboy* uses food to figure the transnational (a Japanese sushi restaurant in Korea) and the flattening of time and space. Even the Korean *gunmandu* dumplings monotonously consumed in his captivity are Dae-su's chief memory of a time in which South Korea "outside" was rapidly democratised and became a global economy (Kim 2011: 194).

Oldboy won the Grand Prix at the 2004 Cannes Film Festival, underlining its transnational success in auteur terms. Park's 2016 film *The Handmaiden* was also selected to compete for this prize, once more underlining the global arthouse credentials of his work. *The Handmaiden* continues Park's use of the cephalopod in the *mise en scène* of the elaborate mansion in which most of the film takes part, owned by a wealthy Korean libertine (Cho Jin-woong) who has prospered during Japanese colonial rule. This includes an octopus in a tank in the basement of the house, which serves as a sort of torture chamber; as well as prominent allusions to the famous woodcut by Hokusai sometimes referred to as *The Dream of the Fisherman's Wife*, which notoriously depicts a woman coupling with an octopus. We might say that this artwork has attained a certain canonical status in relation to the arthouse cephalopod (Carbone 2018). In fact, this points to a stereotyped construction of Japanese culture as inherently perverse that is also present in the reception of certain notorious anime examples – but this takes us beyond the scope of this short article.

For now, we can note that this film contrasts the allegedly "perverse" attraction for the "erotic–grotesque" elements of that Japanese culture, with the attraction between the female Korean protagonists (played by Kim Min-hee and Kim Tae-ri) who reject patriarchal and libertine designs upon them, even if their own relationship is also presented as erotic spectacle. As Suk Koo Rhee notes, the film is highly transnational insofar as "*The Handmaiden* adapts a contemporary British novel about Victorian England [Sarah Waters's (2002) *Fingersmith*] to engage with the issue of modernity in Japanese-occupied Korea" (Rhee 2020: 117). Once more, the presence of the molluscs alludes to the border-crossing nature of nationality, genre and sexuality in the diegesis of these films.[4]

---

4   Had I more space, I would make the case for Bong Joon-ho's (2006) *The Host* as another film to analyse from this perspective – the aquatic monster's tail in this film is

We can finish by alluding to another liminal aspect of these films, between arthouse/auteur cinema, on the one hand, and popular genre or exploitation cinema, on the other. There are interesting films that might be classed as horror but that are sufficiently on the edge of this genre and formally resembling art cinemas to provide additional examples for us here. This locates them in what Steve Rose (2017) has called "post-horror." Space unfortunately precludes me considering whether various H. P. Lovecraft-inspired films would qualify! I will therefore just mention Justin Benson and Aaron Moorhead's *Spring* (2014), with its shape-changing, occasionally cephalopod female lead (Wikipedia winningly describes the film as "romantic body horror"); and Lucile Hadžihalilović's *Évolution* (2015), with its archaic mothers with suckers on their backs. *Spring* premiered at the Toronto film festival and combines downbeat indie style with unorthodox romance between human and shape-changer; the eerie *Évolution* won prizes at the Stockholm and San Sebastián International Film Festivals. One waits with interest to see whether *My Octopus Teacher* will spawn yet more examples of this award-winning, globally circulating and dubiously *enjoyable* trope.

---

sufficiently prehensile to act as a sort of tentacle if one stretches the definition, even if its relation to humans appears more predatory and sadistic than erotic.

Part V

Human Incursions and Environmental Responses

Jimmy Packham

The *Daedalus* and the Great Sea Serpent (1848)

On 6 August 1848, the captain and crew of HMS *Daedalus* saw a sea serpent. The ship was sailing on its passage home from the East Indies and the creature was seen swimming in the waters of the South Atlantic, south-west of St Helena and several hundred miles off the coast of what is now Namibia. The encounter was reported to the Admiralty by Captain Peter M'Quhae and a short account of it appeared in *The Times* on 10 October. A few days later, the newspaper printed the more extensive report M'Quhae had forwarded to his superiors. The creature, we learn from these reports, was in length 60 feet or longer and its diameter "was about 15 or 16 inches behind the head, which was, without any doubt, that of a snake"; it was coloured "a dark brown, with yellowish white about the throat" and it exhibited "no fins, but something like the mane of a horse, or rather a bunch of seaweed, washed about its back" (*Times* 1848b: 3). Sketches of the serpent appeared on 28 October in the *Illustrated London News* (see Figures 33 and 34), which collated all the major published documents relating to this creature to date alongside a number of other notable reports of sea serpent sightings from earlier in the century (*Illustrated London News* 1848: 264–6).

The encounter between the *Daedalus* and the great sea serpent – the name it was given in reports and which tied it into a history of human encounters with monstrous denizens of the deep purported to be "the" great sea serpent – is perhaps the most famous such event of the nineteenth century.[1] Sea serpents certainly have a long and notorious history – from antiquity to the

---

[1]   The *Daedalus*' major rival is the "Gloucester sea serpent" seen off the coast of Massachusetts, most notably between 1817 and 1820 (O'Neill 2003; Soini 2010).

Figure 33. M'Quhae's great sea serpent. *Illustrated London News*, 28 October 1848.

Figure 34. The crew of the Daedalus and the serpent. *Illustrated London News*, 28 October 1848.

famous accounts by Pontoppidan in the eighteenth century and the several major sightings of the nineteenth century, a genealogy traced in numerous Victorian publications.[2] But the story of the *Daedalus* offers a particularly compelling starting point for considering the significance of sea serpents in the nineteenth century for several reasons. First, it prompted a very public discourse, especially in the pages of *The Times*, centred on the probability of a sea serpent's existence – and, moreover, how exactly one might account for their persistence in historical and scientific records, if such things were, in the end, little more than the lurid yarns of unreliable old salts. Second, the *Daedalus* sighting had a substantial cultural afterlife and we can see its influence rippling through newspaper stories, science writing, reference books, fiction, cartoons and satire across the latter half of the 1800s. Swimming through the waters of the nineteenth century, the great sea serpent provides a lens to consider human relations with, and abilities to comprehend, the deep sea; the role of technology in making sense of – or conquering and colonising – the oceanic world; and emergent and competing forms of discourse and knowledge, felt particularly keenly in the tensions between maritime experience and scientific reasoning.

The conspicuous presence of the great sea serpent in the nineteenth-century imagination corresponds to the greater work undertaken during this period of fathoming – literally and figuratively – the largely unknown depths of the ocean. The ocean was also a vital conduit and network for Victorian imperial expansion and an important part of national economies. It makes sense, then, that serpent sightings by British vessels should occur frequently along imperial routes: Captain Harrington was "glad to confirm [the] statement" of the serpent given by M'Quhae, when, sailing back from Bombay aboard the *Castilian*, he believed he saw the same creature (*Times* 1858). Further, from mid-century, the mapping of the seas and seabed became a major scientific endeavour and the commercial work of laying transatlantic telegraph cables began in the 1840s. "The story of the oceans at midcentury," argues Helen Rozwadowski, is "the tale of the expanding human imagining of what the 'deep sea' might be"; and for the "ocean-oriented nation" of Great Britain, "the deep ocean was an important place and a natural site for the exercise of British military, technological, and scientific power" (Rozwadowski 2005: 5–6). The

2   Lee (1883); Oudemans (1892); *Encyclopaedia Britannica* (1886: 608–10).

great sea serpent is a reminder that, in parallel with maritime imperialism and colonisation and as Britain sought to dominate the seas, much that might be strange and wonderful still resided beneath the ocean's surface. Indeed, as Natascha Adamowsky notes, the increased understanding of the ocean and its inhabitants, past and present, did not serve to diminish but rather augmented the sense that the deep sea was full of marvels (Adamowsky 2016: 21–6).

Efforts to imagine or describe the sea serpent help us illuminate such aspects of its cultural life further. The sea serpent, along with its close companion in the cryptozoological imagination, the Kraken, comes to us not just as a strange figure from the deep but as an embodiment of the deep itself.[3] That is, such creatures represent efforts to put comprehensible shape onto the half-glimpsed or elusive qualities of a space largely inaccessible to humanity. Yet, at the same time, the material, bodily form of these creatures – itself so often half-glimpsed and elusive – suggests the difficulties of the human imagination in finding an appropriate means by which the deep might be given shape. It is little wonder, then, that both nineteenth-century and contemporary literature seeking to "explain" serpents should proffer such a disparate array of solutions, including: seals, seaweed, driftwood, a big fish, a row of sharks or whales, prehistoric reptiles, giant calamari, or a whale's penis.[4]

Other depictions of the serpent emphasise the creature's relationship with conceptions of time and technologies. If suggestions that the great sea serpent seen by the *Daedalus* conformed to a common idea that it might be "allied to the gigantic Saurians, hitherto believed only to exist in the fossil state" (*Times* 1848d: 3) this is in keeping with the ways in which geological and palaeontological discoveries of the Victorian period helped "generat[e] speculation that the sea might still contain primordial life" (Rozwadowski 2018: 99). This, in turn, is indicative of an enduring – and ecologically problematic – vision of the sea as somehow outside of time or beyond time's malign influence: following the medieval cartographic practice of placing fantastical creatures at the limits of the known world, here *still* be dragons. In Rudyard Kipling's "The Deep-Sea Cables" (1893) versions of the serpent – "the blind white sea-snakes" – swim simultaneously through the deep sea and deep time, on the seabed, "in the

---

3   For discussion of the strange figurations of the deep, particularly in Tennyson's "The Kraken" (1830), see Packham and Punter (2017).

4   See Oudesman (1892) and Paxton and Naish (2019).

Figure 35. Kipling's transatlantic serpent, illustrated by W. Heath Robinson, 1909.

womb of the world" where in "the ultimate slime" lie "the timeless Things" (Kipling 1909). These white sea-snakes are both Coleridgean in ancestry and partake of an antediluvianism shared with the great sea serpent.

Kipling's poem, too, elides the serpentine with the technological: it is these symbols of Victorian modernity, the transatlantic telegraph cables – from whose perspective the verse is uttered – that, by enabling rapid communication across the Atlantic, have collapsed distance and time itself. Illustrations accompanying the 1909 edition of this poem (see Figure 35) make explicit the formal resonance between the serpentine cable and the cablelike snakes.[5] Both this poem and its accompanying images suggest that the marine serpent-like

5   See also, Andersen (1983).

creatures, via their association with new forms of technology, are at once representative figures of the very ancient and the very modern. And by looking elsewhere we see how technology itself might be figured as that which finally captures, once and for all, the slippery sea serpent.

In his expansive 1892 compendium, *The Great Sea-Serpent: An Historical and Critical Treatise*, the Dutch zoologist A. C. Oudemans offers some compelling reflections on the technological in the hunt for the serpent. The book is a culmination of nineteenth-century interest in its titular creature, gathering together several hundred pages worth of accounts of the serpent, running from 1500 to the late nineteenth century, offering roughly thirty pages of material on the *Daedalus* sighting. At the same time, *The Great Sea-Serpent* provides an analysis of its subject – of whose existence Oudemans is confident – that seeks to fit it into established zoological taxonomies: "I firmly believe that it belongs to the Order of *Pinnipedia*"[6] (Oudemans 1892: 546). It is in a brief and curious note preceding the preface that Oudemans writes: "Voyagers and sportsmen conversant with photography are requested to take the instantaneous photograph of the animal: this alone will convince zoologists, while all their reports and pencil-drawings will be received with a shrug of the shoulders." He further notes that "[t]he *only* manner to kill one *instantly* will be by means of *explosive* balls, or by harpoons loaded with nitro-glycerine" – this approach, he admits, will likely cause the body to sink, and so "the harpooning of it will probably be more successful" (Oudemans 1892: vii). Technology will explode not merely the serpent itself but its myth, too: photography, incontrovertibly providing a reliable, truthful, "instantaneous" representation of things, will settle the matter. Oudemans was prescient in some respects, insofar as photography and film have provided some of the most iconic images of a plethora of cryptozoological creatures, from sea serpents to the grainy "surgeon's photograph" of the Loch Ness monster and the minute-long "Patterson-Gimlin" footage of Bigfoot. His faith that such technology would provide transparent and self-evident proof of cryptids has not, however, been so rewarded.

The last point to take away from Oudemans' remarks is the way in which he frames this pursuit of the great sea serpent as playing out between "voyagers and sportsmen" on the one hand and "zoologists" on the other. Oudemans'

6   A taxonomic order comprising seals, sea lions, and walruses.

own career as a zoologist notwithstanding, to see the dispute over the serpent's existence in this manner is to draw attention to different orders of knowledge and experience, different ways of seeing the world. It is also to situate the great sea serpent as an evocative focal point in considerations of the professional-isation of science and the emergence, across the latter half of the century, of "science as an authoritative and distinctive cultural discourse" (Holmes and Ruston, 2017: 8; see also Beer 1996 and Ellis 2014).[7] The serpent prompted its audience to reflect variously on evolutionary biology, biological classification, geology, the earth's antiquity and the fossil record. Again, the *Daedalus* sighting and its history of reportage is instructive in this respect, especially the letter printed in *The Times* on 14 November 1848 from the prominent biologist and palaeontologist, Richard Owen – whose intervention in the *Daedalus* case was surely fundamental in ensuring its longevity in the popular imagination.

Owen pens his letter, it seems, out of simple exasperation: "as I continue to receive many applications for my opinion of the 'Great Sea Serpent,' I am desirous to give it once for all through the medium of your columns" he writes to *The Times*' editor. Owen argues that what M'Quhae saw was likely a large wayward seal, specifically "Anson's sea lion, or that known to the southern whalers by the name of the 'Sea Elephant,' the *phoca preboscidia*" (*Times* 1848e: 8). Owen's conviction that sea serpents do not exist is, as he explains, rooted above all in the facts that no genuine sea serpent bodies or bones have ever been recovered and that their remains are completely absent from the fossil record. We are deceived, he says, if we think we have seen a sea serpent, noting with a final flourish:

> I regard the negative evidence from the utter absence of any of the recent remains of great sea serpents, krakens, or Enaliosauria, as stronger against their actual existence than the positive statements which have hitherto weighed with the public mind in favour of their existence. A larger body of evidence from eye witnesses might be got together in proof of ghosts than of the sea serpent. (*Supplement*, 1848)

In a pointed reply, M'Quhae responded to "the animadversions of 'Professor Owen.'" All who saw it, he notes, "are too well accustomed to judge of lengths and breadths in the sea to mistake a real substance and an actual

---

7   On the changing face of biological sciences across the nineteenth century, see Holmes (2017).

living body" for something else, and his statements will be left "as data where-
upon the learned and scientific may exercise the 'pleasures of imagination' until
some more fortunate opportunity shall occur of making a closer acquaintance
with the 'great unknown' – in the present instance most assuredly no ghost"
(*Times* 1848f: 8).

It would be misleading to suggest that Owen's is a simple index of scientific hostility towards popular belief: prominent scientists and contemporaries
of Owen including Louis Agassiz and Charles Lyell expressed varying forms
of support and enthusiasm for the reality of sea serpents (Lyons 2009: 17–50;
O'Neill 2003: 117–18). What matters, however, is the way in which Owen
deploys his authority as a naturalist and palaeontologist, and the technical
vocabulary of his disciplines, against M'Quhae's substantial but ultimately
unscientific maritime experience. In quest of proof, Owen need "not ask for
the entire carcase" of a sea serpent, for "[t]he structure of the back bone of the
serpent tribe is so peculiar, that a single vertebra would suffice to determine
the existence of the hypothetical Ophidian" (*Supplement* 1848). Owen's authority is particularly emphasised as he signals his skilled ability to reconstruct
the whole from a fragment.

In turn, M'Quhae holds Owen's technical terms at arm's length – placing
them in quotation marks – while emphasising the precise and methodical
character of his own observatory powers and his expertise interpreting a sea-
scape that he, not Owen, had navigated. M'Quhae insists on the deliberate
nature of his own practice: the crew "duly considered and debated" the creature before arriving at their conclusions (*Times* 1848f: 8). The captain's reply
also embodies a version of what Hester Blum terms the "sea eye," an "experi-
ential vision" peculiar to "the sailor's labor knowledge" and "special vantage
point in an environment inaccessible to most people" (Blum 2008: 116). For
Blum the sea eye, uniting maritime labour and processes of contemplation,
is an important foundation for the remarkable cultural work produced by
sailors in the nineteenth century. Here, we might see it adapted to slightly
different ends: as a means to insist on a specialist form of oceanic knowledge
and comprehension.

Punch magazine helped cement Owen's reputation as the "serpent killer"
when it published a pithy verse summarising this exchange under the title "More
Last Words" (*Punch* 1848: 243) – a wry allusion to *Times* reports appearing

under headings like "The Great Sea Serpent Again" (*Times* 1848c: 6) and "The Sea Serpent Once More" (*Times* 1848f: 8). The *Daedalus'* presence in *Punch* introduces the final point to be made here, too – that the sea serpent, in the latter half of the century, figured as a shorthand for silly season, the period of the year in which, with Parliament not sitting, the news was often seen to be filled with trifles or farcical stories.[8] Jules Verne burlesques both silly season and the "interminable argument [...] between the credulous and incredulous" on the matter of sea monsters in the first chapter of *Twenty Thousand Leagues Under the Seas* (1869–70). When global news outlets remain baffled by a submarine presence that turns out to be Captain Nemo's *Nautilus*, "[e]ach imaginary gigantic creature," we are told, "resurfaced in the papers, admittedly short of good copy." Here, the *Daedalus'* serpent – evoked via Verne's reference to "Captain Harrington" of "the *Castilian*" – takes its place among a coterie of marine monsters, including Moby Dick, the Kraken and the creatures reported by Pliny, Aristotle and Pontoppidan (Verne 1998: 7), thereby inhabiting a lineage that is purely fictional and notionally nonfictional. The *Daedalus* sighting and its afterlife speaks finally, then, to an enduring ambivalence towards oceanic depths and the epistemological difficulties involved in coming to know and reliably explicate some of its seemingly monstrous – or monstrously imagined – realities.

8   For first introducing me to this usage, at the 2021 "Haunted Shores" symposium, I am indebted to Charles Paxton.

Lauren A. Mitchell

The Lure [Córki dancingu]
(Agnieszka Smoczyńska. 2015)

Of all mythical creatures, the semiotics of the mermaid are uniquely ro-manticised: sanguine flowing movements, mysterious powers, a "siren song" that lures men towards them and a way of being unknowable that draws out human curiosity. She is, in the words of Lori Yamato, "a creature that is both inherently less than human and yet all too human" (2017: 300). Of this world and also not of the landside of the world, she has been a persistent object of fascination while, at the same time, inhabiting a body that is discordant with the expectations of the human world. Katie Noson describes, "the ambiguous morphology of the siren-mermaid reflects both an apprehension regarding the disabled female body – itself often coded as 'monstrous' or grotesque – as well as an attempt to reconfigure that body as desirable and capable" (2016: 18).

Agnieszka Smoczyńska 2015 film, *The Lure* engages with the concept of the mermaid and disability through its engagement with performance and display culture, which are relevant to a number of contemporary disabled performance artists as well as the complicated historical precedents put forth through the medical community. The film pivots Hans Christian Andersen's *The Little Mermaid* into a story about the mermaid as a rock-star celebrity icon, imbuing the fairy tale with the hypervisibility of performance and dis-play culture that was a major facet of European medicine in the eighteenth and nineteenth centuries, particularly surrounding bodies that were deemed "othered," monstrous and fascinating in equal proportion. This is a unique contrast to the original story, where the titular Little Mermaid has to radically revise her body at a great emotional and physical expense to herself in order to blend into human society. This assimilation renders herself, her pain and her intentions invisible. The story ends with a point of supreme invisibility; when the prince marries someone else, the little mermaid turns into sea foam. In failing to secure his love, and therefore in failing to be seen and regarded by him in a particular way, her life ceases to exist.

Only selectively faithful to the fairy tale, Smoczyńska's film repurposes the story in a Polish nightclub where two mermaid sisters, Silver (Marta Mazurek) and Golden (Michalina Olszanska), become a featured act as much for their music as for the appeal of their phenomenal bodies. The "monstrousness" of the sisters is not due to their physical forms, but rather because their engagement with the human world is primarily to satisfy their hunger. Their beautiful singing voices are part of their siren-song in order to attract human men to kill and eat. They choose to stay on land because one of the sisters, Silver, falls in love with a young human man. The film confronts the question of disability in an interesting, oblique way by routing it through questions of the historical underpinnings of spectacle and performance, and where the concept of "the mermaid" confuses binary understandings of monstrous and desirable bodies.

Its setting is a nightclub that is part music venue, part restaurant, part strip club, where burlesque performers are billed alongside contortionists. It is a place where unusual forms and norms of embodiment are featured as spectacular not only because of their magnificence, but because they are phenomenal, entertaining, *spectacles* with fantastic physical structures. Unlike the fairytale of the little mermaid that dictates that she must blend into human society, Silver and Golden are celebrated for their talent and because of their mermaidness. But despite their catapult into local icon status, they are only celebrated inside of the context of the show they put on and are otherwise kept hidden; an echo of the heritage between the burlesque performances that constitute *The Lure* and the freak show.

The Lure is a film that structures the mermaid archetype in dual ways that connect to the fairy tale, as well as to the biological fascination with otherness that was so prevalent in the eighteenth and nineteenth centuries, when physiogonomic practices, anatomy and the Darwinian question of what defines a human was a major cultural focus in stories as well as in science. Additionally, the figure of the mermaid has found its place in the display culture of anatomical museums by way of a rare congenital condition called "Sirenomelia," in which foetal leg bones are fused in utero, forcing the feet to point outward in the appearance of a fin. It is a condition that is typically considered, "incompatible with life" due to its association with organ malformation, with a few rare exception (see Fadhlaoui et al. 2010; Goodlow et al. 1988; Romano et al. 2006). As with spina bifida and other open neural tube issues, the fact that post-birth survival rates are so low have amplified the availability of "mermaid fetuses" and their skeletons in pathological museums, freak shows and other sites of "wonder" among people with normative bodies. Sirenomelia has frequently been on the roster as a subject of study in medicine within two of its core genres: the pathological museum and its written corollary, the case report.

The subject of displayed bodies often brings up warranted concerns regarding power structures and the hierarchies inherent to who gets to look, and who must tolerate being looked at; this gets reinforced by the fact that the aforementioned genres of medical study are created "by and about medical professionals" that are then shared with the public in an authoritative, educational capacity (Anderson and O'Sullivan 2010). Disabled people, particularly those with visible disabilities have a multifaceted relationship with the medical visual culture that constituted the freak show and the medical museum that would render them into a series of body parts and pathological terms. The lingering nineteenth-century fascination with "otherness" and the morphology that made humans human was triangulated through Victorian Britain's imperialist project and the Anthropological Shows of London. Medicine and science were a part of the spectacle of these shows, which were also integrated into explanations for the lay-public (Durbach 2009: 20–37). Some were anatomists who wanted to lay claim on disabled bodies in an attempt to culminate their own forms of celebrity status. The trade-off would be that this is how many of the performers were able to receive medical care, though in its most exploitative form. And some of these bodies would, invariably, continue their performances post-mortem, on display shelves.

Without attempting to restore this deeply problematic history, it is also important to note how disabled performance has evolved in order to think about how representations such as *The Lure* could be seen as potential reclamation. Historian Nadja Durbach has noted the discordance between the public understanding of the "freak show" from its nineteenth-century origins to now. She comments that the word "freak" was a performance identity and argues that it was not necessarily a bad one – indeed, that it was a means by which to achieve a celebrity status by way of their spectacular bodies. This similarly troubles our current perspective of what it means to be "able-bodied." She writes that the term "able-bodied [...] was not rooted in the deformity of the body per se but rather in its capacity for labor" (2009: 18). In other words, because such performers were able to make an income precisely due to their bodies, it made them, by Victorian standards, very abled – if only from an economic perspective.[1]

Disabled writer and performer Eli Clare writes:

> I relish the knowledge that there have been people who have taken advantage of white people's and nondisabled people's urge to gawk. I love that disabled people at one time were paid to flaunt and exaggerate their disabilities. At the same time, I hate how the freak show reinforced the damaging lies about disabled people and nondisabled people of color. (2015: 95)

Indeed, many disabled artists "stare back" at this framework to use their bodies in creative, empowered ways in resistance to the "trauma porn" that constitutes much of their documented history. Noson writes of the "ambivalence expressed by disabled authors regarding their disabilities – as a source of pride but also as the cause of repeated discrimination and an obstacle to full participation and inclusion in many of the aspects that comprise a rich human life" (2016: 18). Clare zeroes in on the complicating elements of performance; one of the core values of contemporary performance art is consent such that the artist, at the centre of the room, is the one "calling the shots," but because a given performance is a co-constructed display between performer, audience and a potential slew of producers behind them, there are always diaphanous layers that can shift power if they are not regarded carefully.

In *The Lure*, disabled performances remain an object of fascination through "burlesque culture," something that is absolutely shared with current performance artists – namely, Mat Fraser, a performer with phocomelia (shortened arms) incorporates sexy dance performances into his shows *because*

---

1   David Gerber in, "The 'Careers' of People Exhibited in Freak Shows: The Problem of Volition and Valorization," emphatically contrasts Durbach's point. I would be remiss if I did not underscore that I do not view Durbach's point as a way to excuse the exploitation of the freak show culture, but rather to unpack the labour aspect of such performances and the bizarre form of "celebrity" that followed.

of, not in spite of, his disability, as well as the performance group Sins Invalid, who attempt to challenge norms of sexuality and embodiment through their performances (see also Chemers 2003 and Stephens 2005). Though the larger metaphor of the mermaid sisters having to assimilate into a world that is not built for them is relevant in *The Lure*, it is the discordance between their "freakish" performance identity and their desire for actual life that amplifies an implicit narrative of disability.

The Lure begins at a shoreline, where Silver and Golden look at a family that comprises the nightclub band, "Fig'n'Dates." Silver's gaze softens at the sight of their young son. When her sister joins her at the surface, they begin their siren song: "Help us to the shore/you have no need to fear, my dear/ we won't *eat you/eat you/eat you.*" Hypnotised, the son (Mietek (Jakob Gierszal), credited as "the Bassist") and his father (known as "the Percussionist") step towards the water, and the scene cuts midway through the family matriarch's scream. Seamlessly, the scream transposes onto a close-up of family matriarch singing Donna Summers' "I Feel Love" in the nightclub where the family band plays, and where the mermaid sisters will be adopted into the motley crew of entertainers. This entrée into the performance world begins with an impromptu physical examination of the girls, resonant with the symbiotic culture and inappropriate overlap of the freak show and the medical community. To the tune of "I feel love," the manager follows a pungent, fishy smell to where the girls are hiding backstage, in the green room. From the looks of them, they are two nude young women who seem too young to be in a nightclub. It is with the mermaids' shape-shifting ability that, admittedly, *The Lure* skirts certain questions of disability and societies that can only accommodate for typical body structures.

As the manager interrogates the family, the Percussionist shoos his wife and son away so that the girls are alone with only the two men in a dynamic which troubles a perception of who, exactly, is the predator in the room: the carnivorous mermaids, who began the film with false promises that they will not eat men? Or the clownish, older gentlemen who are alone in the room with two nude ingenues? The Percussionist he asks them to bend over, turn around and spread their smooth legs to show how smooth and genital-less they are and demonstrates the shape-shifting capacities of the sisters by pouring a glass of water over their legs which makes them slowly turn into their true form of magnificent, eel-shaped tails. Once they change back into their true form, he guides the manager towards the slit at the bottom of their tails to point out where their sexual organs – the unceremonious "cloaca" – are located. As the men stick their fingers into the slit, the girls coyly giggle.[2] They are told by the club manager that they can sing backup for Fig'n'Dates and "strip," which is enough to make them happy, as they become incorporated into "showbiz" life.

The giggling cuts the edge of the gravitas off of the exam, a move that allows the audience to equally bask in the strangeness of the mermaids. In the diegetic world of the film, it also indicates that the mermaids are not typically the victims among men; that the sexualised elements of their bodies are a conduit to hunting and other goals. In this case, their goals seem to be to enjoy the adventure of staying in the human world, and, for Silver, to spend more time around the boy she falls in love with. This assessment of their wonder, and, more importantly, their ability to pull in a crowd, recalls scenes of medical examination, where Silver and Golden are touched and gazed upon very intimately. Although it would be easy to foreground how this scene could easily perceived as sexual exploitation, what is equally compelling is how this makeshift genital exam in a dressing room harkens back to other examples in history that convey a disproportionate interest in the reproductive organs of women and disabled people. The mechanics of sex has always been a point of curiosity in the human imagination.

In fact, the major way in which the sisters might be "disabled" in the film is through sex and desire. Silver is guided by her love (or, perhaps, lust) for Mietek, even when he tells her, "No offense, but to me, you are a fish, an animal to me," after she tries to initiate sex – despite the fact that they are inseparable for most of the film, that they kiss and exhibit other forms of physical intimacy. The fact that Silver does not have a "normal" human vagina stifles her ability to integrate fully into the human world. At the same time, Golden's seething rebellion towards the limitations of position as a performer in the human world unfurls. The family counts the money they earn from the newly-mermaid-enhanced Fig'n'Dates performance while regarding the girls

2   Anatomical reports and pathological museums show a persistent fascination with genitals, especially of women and marginalized peoples. See also: Mitchell, "Erotic Surgery: J. G. Ballard's *Crash*, Octavia Butler's 'Bloodchild,' and the Visual Legacy of the Medical Museum," forthcoming in *Configurations*.

sarcastically, and then later with fear when it comes out in a news report that Golden had sneaked away to seduce and eat a man. In the wake of the family's fraught silence, Silver becomes angry at the family and the ways in which the girls have been kept hidden: "Why don't we ever go anywhere? [...] All we do is stay in this house," and "Why don't we see any of the money?"

The money is, of course, meaningless to the mermaids; their relationship to the family is not contingent upon any material needs, but rather their emotional attachments to the human world – for Golden, her identity as a performer, a potential desire for revenge and her desire to take care of her sister; for Silver, it's the more obvious desire for love. For this, she subjects herself to a mutilation, a word aptly used by Lori Yamoto to describe the original Little Mermaid story, where she and Mietek seek out a doctor to transplant the bottom half of her body – her tail and the cloaca in question – with a set of human legs and "a real pussy." The scene, set up in a dark, wooden, makeshift operating room, is reminiscent of some of the gothic scenes of early medicine. Silver lies awake, singing, on a bed of ice next to a donor (whom we assume is dead). But her song is abruptly cut off with the loss of her voice once the transplant begins.

History is beleaguered with stories about women's bodies as they are scrutinised and surgically manipulated for a variety of sexual "insufficiencies," whether due to an overflow of desire for sex, its frigidity, or its hysterics. And, like so many of those surgeries, the result for Silver and Mietek is that she continues to be sexually repulsive to him: they try to have sex and fail when her wound bleeds onto him. Their "love" scene ends with a close-up of his face, exasperated. Like the original fairy tale, he moves onto an easier, humanly legible lover whom he marries – and we know how the story ends for Silver. Noson aptly close reads an analogue for this fictional surgery by way of case reports for sirenomelia correction:

> [P]rocedures are undertaken to reconstruct the vagina [...] The imperative to separate the mermaid's legs fuses heteronormative fantasies of controlling the monstrous female body with the normalizing imperatives of medical cure[.] [...]. In other words, the rehabilitation of the "normal" body in medical terms is inexorably bound to a rehabilitation of femininity and sexuality (2016: 24).

In capitulating to a human standard of normalcy, Silver inadvertently agrees to divide herself into a series of disparate, dysfunctional parts in a place that is not, nor will ever be, home to her. In addition to some of the larger impacts of her surgery – namely, the entire removal of her tail, the loss of her voice and subsequently the loss of her performance identity – she also cannot walk well. As she hobbles towards Mietek, she is reminded by a friend that merfolk are meant to be guests on land. Although the mermaids may have been able-bodied in the context of their enfreakment, their income-generation and their shape-shifting ability, her participation in the human world was always contingent upon a freak performance. In this, there is a similar capitalistic paradigm of the original freak shows that reinforces celebrity while diminishing humanity. The tragedy is that Silver can only mimic human normalcy and cannot achieve an appropriate balance of visibility: she is either a too visible, or invisible, with no median option. When she gives up the spectacle of her mermaid body, she turns into foam; into nothing.

Matt Melia

The Meg (Jon Turtletaub, 2018)

Introduction

It's hard not to read *The Meg,* with its depiction of big tech, corporate greed, environmental catastrophe and the exploration, invasion and colonisation of the ocean space as an allegory for voracious consumption and competition – the Megalodon itself as an embodiment of a nature biting back. In this era of neo-liberalism and rampant capitalism there is always a bigger shark (and this film gives us two!). The central underlying preoccupation with consumption also applies in a postmodern sense: the film itself "consumes" a variety of other film texts – from *Jaws* (Spielberg, 1975) to *Jurassic Park* (Spielberg, 1993) to *Star Trek IV: The Journey Home* (Nimoy, 1986) – one of the pair of "pet" whales in *The Meg* is named Gracie, a reference to the pregnant whale at the centre of the *Star Trek* film. In *The Meg* of course, there is no happy end for Gracie the whale, who is literally and metaphorically devoured by the first of the film's two Megalodons. Later in the film, another whale is indiscriminately blown up by the film's real antagonist, the shark-like billionaire, big tech entrepreneur and investor Jack Morris (Rainn Wilson), who like the first Megalodon is also devoured by the second. The film therefore sets up a deliberate parity between Morris and the Meg.

In this chapter I intend to offer a contextual understanding and a reading of *The Meg* in terms of its genre, cinematic and textual positions and a reading of the monstrous prehistoric sharks in *The Meg* as "textual devourers" and embodiments or signifiers of environmental revenge and catastrophe.

Lizards, Apes, Dinosaurs, Sharks – and Megalodons

Environmental Exploitation

I. Q. Hunter positions the film as the latest entry in an increasingly prehistoric subgenre (now nearly fifty years old). In *Cult Film, A Guide to Life* (2016), he notes how in the wake of the success of *Jaws* (1975), film makers keen to cash in on the popularity of the film exploited the gap in the market left in its wake. Films such as *The Deep* (Yates, 1977) (also based on a Peter Benchley novel) combined the aquatic thriller (with its central treasure hunting narrative) with aspects of oceanic horror (one sequence in *The Deep* includes a monstrous moray eel). Hunter organises the films that surfaced in the wake of Spielberg's Great White into "two waves," "Jawsploitation" and "Sharksploitation." The former followed on "the heels of the film's [*Jaws*] release [...] a model exploitation cycle capitalizing on *Jaws*-mania." These films were "parasites" on the back of Spielberg's shark and reworked the plot to "to showcase a variety of killer beasts": Barracuda (*Barracuda*, 1978), Grizzly Bears (*Grizzly*, 1976), Octopi (*Tentacles*, 1977), etc. (86). The second wave he defines as "Sharksploitation", films which "initially either loosely remade *Jaws* [...] or, ignoring the plot of *Jaws* entirely, were simply films with sharks in them" (83). Neil Jackson locates the term within the context of a more recent contemporary cult/exploitation market. These films hybridise genres and forms of popular and cult film (with titles such as *Shark Exorcist* (Farmer, 2015) and *Frankenshark* (Zebub, 2017). "Sharksploitation" he defines as a

> [p]hrase applied to the ongoing spate of direct to DVD/streaming titles [...] a twenty first century home viewing phenomenon which evolved into increasingly more outlandish sci fi/fantasy hybrids such as *Mega Shark* (2009–2015); *Dino Shark* (2010), *Sharktopus* (2010) and the *Sharknado films* (2012–18). (Jackson 2020: 245)

With *The Meg*, he argues, "the cycle appears to have come full circle back to the multiplex following the $385 million international box office gross" (ibid.). *The Meg* appeals to a hybridised audience – not only, (as Hunter points out) does the film play to the cult/exploitation market but also Popular blockbuster audience. Its acts as a vehicle for British action star Jason Statham who brought with him a readymade action cinema audience (the climax of the film sees "the Stath" getting into a fist fight with the gigantic Megalodon).

We may argue that *The Meg* combines, or deliberately consumes, aspects of both *Jawsploitation* and Sharksploitation – especially given its sense of self-awareness, irony and the amount of knowing references to Spielberg's film (and others) throughout; The Megalodon's voracious appetite (especially when it comes to whales and other Megalodon's) becomes a cipher for the film's own habit of devouring other texts.

The film was released two months after the second instalment in the *Jurassic World* franchise, *Jurassic World: Fallen Kingdom* (Bayona, June 2018) which took 1.31 billion USD at the Box office dwarfing *The Meg's* takings of 530.2 million USD. In *Jurassic World* (2015), the first of the extended *Jurassic Park* franchise, a Great White Shark is depicted being fed to a giant Mosasaurus, as part of a spectacular performance for the paying customers of the newly opened theme park (see Figure 36).

Figure 36. Mosasaurus vs. Shark. *Jurassic World*, directed by Colin Trevorrow (Universal Pictures, 2015).

This self-aware statement recognises the displacement of the Shark movie, and its consumption by bigger, more frightening monsters able, with the help of new cinema technologies, to be rendered in a more immediate and spectacular way. By the time of the re-emergence of the *Jurassic Park* franchise, the taste for giant prehistoric monsters had been well and truly whetted, especially with the then also-emergent "Monsterverse" from production house "Legendary." The franchise had begun in 2014 with the reboot of *Godzilla*, directed by Gareth Edwards, and was followed up by *Kong: Skull Island* in 2017 directed by Jordan Vogt-Roberts (since then two other films in the franchise have been released, *Godzilla King of Monsters* (Dougherty, 2019) and *Godzilla vs Kong* (Wingard, 2021) to popular if not critical acclaim). Given the dominance of giant prehistoric Lizards, Gorillas and Dinosaurs, we might ask where the Megalodon had been hiding during this feeding frenzy. By 2018 the shark had increasingly lost ground to these big budget Hollywood behemoths – the legacy of *Jaws* swallowed up in the shallows of parodic exploitation (or "sharksploitation") cinema dominated by the "so bad its good" *Sharknado* franchise (2013–18). There have of course been several attempts to make more commercially profitable shark movies, including among others both *The Reef* (Traucki, 2010), *47 Meters Down* (Roberts, 2017) and *The Shallows* (Collet-Sera, 2016) and more recently *Great White* (Wilson, 2021). Blame for Hollywood's loss of faith in the shark has been widely blamed on the financial failure of *Deep Blue Sea* (Harlin, 1999)[1] which as Neil Jackson notes, opened the hatch for the plethora of straight to video parodic and genre hybridising Sharksploitation films that followed. Max Covill reflects that the failure of Deep *Blue Sea* (as well as changes in studio management) directly impacted the production and release of *The Meg*, which prior to its 2018 release had been stuck in development hell for twenty years:

> Both *Deep Blue Sea* and *The Meg* were being developed by rival studios at the same time. Warner Bros. released *Deep Blue Sea* in theatres in the summer of 1999 and the film was a disappointment. *Deep Blue Sea* also shared a lot of DNA with *The Meg* being they were both shark attack movies. Producers deemed that audiences just weren't into shark

1   *Open Water* (Kentis, 2003) evaded the financial fate of most post-*Deep Blue Sea* shark movies by billing itself as a survival film first and foremost, which just happened to have sharks in (although a shark's fin is prominently visible in the film's marketing).

Figure 37. A bigger fish. *The Meg*, Jon Turtletaub (Warner Bros, 2018).

movies at the moment and *Deep Blue Sea* had a major role in the lack of production of *The Meg*. (Covill 2018)

The Meg attempts to return the shark movie to the upper echelon of the Monster movie genre, staking its claim in this big-budget arena of giant pre-historic monster movies by introducing not one but two Megalodons – the second arriving out of the blue (so to speak) later in the film to devour the carcass of the first (see Figure 37), not only establishing the film's position in this increasingly crowded market of giant monsters, but also ironically sinking its teeth into the subgenre from which it emerged.

Here the film also offers further layers of self-referentiality and intertext-uality: Jonas' realisation that the size of the teeth on the carcass of Meg #1 don't match the bite marks on the research station, recalls the sequence in *Jaws* in which Hooper (Richard Dreyfuss) tells the fishermen that the shark they've just caught isn't "*the*" shark, and pre-empts the enormous second Megalodon leaping from the waves to devour the first. This (textual) devouring is also made clear in the film's marketing (see Figure 38).

Kent Hill notes: "While Spielberg's film is by its nature a more intimate piece; the shark menaces a small community and finally three men set out to kill the beast, MEG is something we are definitely going to need a bigger boat for" (2018). *The Meg* is textually aware of *Jaws'* impact on the shark movie setting itself up as a bigger, more spectacular successor, as coded by the second giant Megalodon devouring the first. At the end of the film the second Megalodon

Figure 38. *The Meg* film poster (Warner Bros. Pictures, 2018).

is itself ironically devoured in a feeding frenzy of tiny parasites (actual sized Great Whites!), emphasising and anticipating the place of *The Meg* in a continuous cinematic food chain.[2]

2   In 2015 horror director Eli Roth (*Hostel*) became associated with the film and given the amount of shark cannibalism in the film we are left to wonder how cannibal exploitation movie aficionado Roth might had realised it.

Unlike the other recently resurrected prehistoric beasts which look to the Japanese Kaiju films, or King Kong (Hollywood's original mega monster), if we differentiate The Megalodon from the Great White Shark (considering it as a monster in its own right) it has no major cinematic precedent. In fact, in his article for the *Daily Jaws* from around the time of the film's release, Dean Newman is careful to differentiate between Great White and Meg saying, "Contrary to popular belief, the Great White and The Meg are not closely related" (Newman 2018). But if *Jaws* is a clear reference point for *The Meg*, then in *Jaws* the film clearly aligns the shark with the Megalodon:

> The Megalodon has a blink and you'll miss it cameo appearance in *Jaws* when Chief Brody is flicking through shark books, it's that iconic image we all know and love of two rows of Victorian men stood in its jaws. (Newman 2018)

The film might also be read as a response by Hollywood not only to the *Jurassic World* franchise but also to the earlier low budget Sharksploitation films in which the monstrous prehistoric ancestor of the Great White Shark played a central role, popular interest in the Megalodon was, perhaps, provoked by the 2013 annual *Shark Week* on the Discovery Channel. The sensationalist documentary *Megalodon: The Monster Shark Lives* (Glover, 4 August 2013), as Vincent Campbell notes, *cemented* the popularity of these prehistoric monsters, and *spawned* a whole array of low budget, parodic Meg films, including *Raiders of The Lost Shark*; *Jurassic Shark*; *Shark Attack 3: Megalodon*; and *Megalodon*, attracting five million viewers with its "account of Megalodons alive in the present day" (2020). Campbell reflects that the pull of the *Jaws* narrative, of the dramatic, scary and dangerous shark persists and extends into the subjunctive imaginings of what a Megalodon could do to us, if it were alive today, freed from the constraints of scientific verisimilitude of the conventional documentary (ibid.).

The Mariana Trench and *The Meg*

The film's action takes place in and around The Mariana Trench, the deepest location on earth measuring over 1,000 miles long and 36,201 feet (7 miles) deep. *The Meg* was, of course, not the first film to utilise the trench as a space of the unknown and eldritch. James Cameron's documentary *Deepsea Challenge* (2014) took viewers on a three-dimensional journey into the depths of the Challenger Deep (the deepest point of the trench). *Deep Rising* (Sommer, 1998) dealt with monstrous cephalopods rising from the depths of the trench, and more recently *Underwater* (Eubank, 2020) used the Trench as the resting place of HP Lovecraft's Cthulhu and the Deep Ones. Nevertheless, as a location, it has been used, perhaps surprisingly, relatively infrequently.

US Navy lieutenant Don Walsh and Swiss Engineer Jacques Picard were the first to reach the bottom of the trench as far back as 23 January 1960. In 2012 film director and oceanic documentarian James Cameron's dove 35, 787 feet to the bottom of the trench and in 2019 Victor Vescovo, a retired naval officer descended 35, 853 feet to the bottom. Narratively however, *The Meg* ignores this history and posits a fictitious thermal self or "Thermocline," a layer of cold water which in the films narrative has been presumed to be the bottom of the trench or "Challenger Deep." The film adds another narrative layer of mystery to the trench by offering unexplored depths beneath this thermal shelf through which the Megalodon surfaces as a result of unwanted human intrusion. The unexplored Ocean depths provide a space as alien, unsettling and unknown as outer space – yet an also an untouched underwater Eden (see Figure 39).

The Megalodon is the guardian of this uncolonised, underwater realm – at first facing off against a giant squid attacking a rescue glider, then the intrusive sub itself. The thermal shelf presents a barrier between a colonised and an uncolonised space: above it the man made, technological palace of the half-submerged research station paid for and funded by the shark-like billionaire Jack Morris – who has no actual idea what it really does, nor does he really care as long as it makes money for him and looks cool – an embodiment of capitalist exploitation and hubris (and a parody of big tech hipsters like Mark

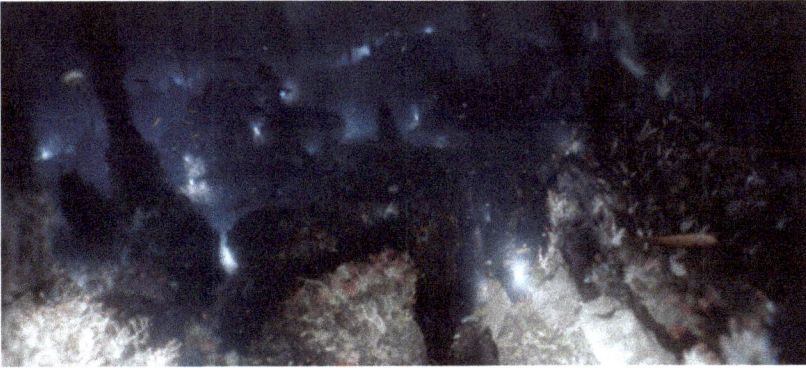

Figure 39. Below the Thermocline. *The Meg*, directed by Jon Turtletaub (Warner Bros, 2018).

Zuckerberg, Steve Jobs or even Elon Musk). The films excursion beneath this "thermocline" depicts a rich if mysterious new world, as yet untouched by the ecological decimation above: a flourishing abundant world of alien sea life. The reality is more prosaic and tragic. When Vescovo descended in 2019 (the year after the film's release) he was to find a plastic bag lying on the floor of the trench.

The Megalodon is used, particularly in the film's second half as a signifier of global ecological disaster (as Newman indicates the real-life prehistoric Megalodon was likely killed off by climate change – making it an apt signifier of global environmental disaster). In an article for the Guardian from after the film's release, Rafael Motamayor offers a reading of the film suggesting that the film "takes the Megalodon's side" (2018). In one sequence the discovery of a decimated school of sharks proves not to be as a result of the Meg but shark poachers. One of the character's quips that the Meg has been getting its own back, before the severed arm of a poacher is removed from the water. Motamayor writes:

> In 2017 there were 155 incidents of shark-human "interaction" (five fatal), while 100 million sharks are killed each year by humans. Despite this being a film about a predator that must be destroyed before it devours too many humans, it wants to say something important about our real-life treatment of sharks and the environment. (2018)

The treatment of sharks in film as engines of death remains problematic. In the wake of an increase in shark killings, post-*Jaws,* author Peter Benchley reflected he would have thought twice about writing the book. The more recent film, *Great White,* presents a sequence in which the lead character graphically knifes a shark to death while riding on its back. *The Meg* is more circumspect in its representation of the shark/megalodon and its approach to the current global environmental crisis. One of the research scientists Jaxx (Ruby Rose) is also an ex-environmental (Greenpeace) activist and the film draws a distinction between the supposedly benign aims of the scientists and Morris' neo-liberalist greed. The Megalodon however makes no such ideological distinction. All are intruders in its realm, and despite their intentions, it is hubris and human entitlement which leaves the way open for it to surface above.

It is during the climactic the final sequence in which the Megalodon attacks a crowded beach in Sanya Bay, China, where the film sets out its most stringent critique. We are first presented with an entitled and wealthy wedding party celebrating aboard a luxury yacht and frolicking in the ocean, with little regard to the ocean space around them. The frivolous partying above is countered by a subsequent underwater shot of the beast swimming through discarded plastic bottles, plastic tubes of sunscreen, etc.

The Megalodon's next port of call is to attack the shore, which is packed with plastic lilos, zorb balls and rubber rings floating on the surface of the water (see Figure 40) with an overhead shot of the monstrous shape of the Meg beneath. It is interesting that given the film was a US-China co-production, that this feels like a veiled criticism of both countries' environmental footprint: China is the biggest global producer of plastics and in 2018 accounted for a third of the world's plastic production. In the same year the United States produced thirty-seven million tonnes of plastic. Both the United States and China came in at number 1 and 2 as the "wealthiest countries in 2019." As it launches its attack on the bathers, The Megalodon, the Leviathan from the

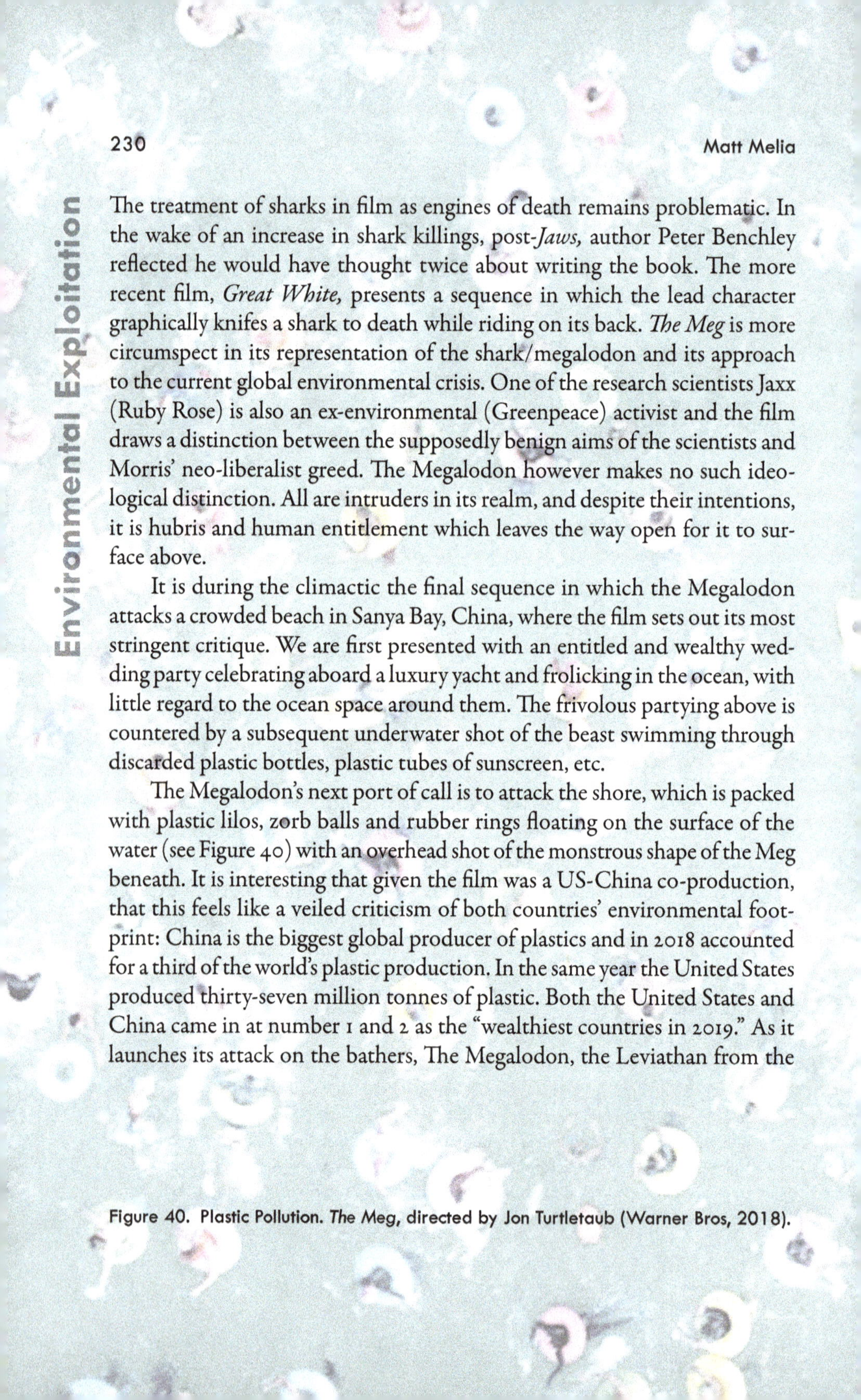

Figure 40. Plastic Pollution. *The Meg,* directed by Jon Turtletaub (Warner Bros, 2018).

deep, becomes a living embodiment or harbinger of the inevitable climate catastrophe which stems from capitalist, neo-liberal hubris, over consumption, decimation and colonisation of natural ocean resources, climate change and plastic pollution and the monster, like any good film monster, reminds us of our personal responsibility.

Environmental Exploitation

Jennifer K. Cox

Into the Drowning Deep
(Mira Grant, 2017)

Under pen name Mira Grant, the Campbell, Hugo and Locus Award-winning author Seanan McGuire animates folk and fairy tale elements with unexpected and disturbingly realistic detail, regardless of genre. Her texts often place fantastic creatures in real-world settings and the disastrous results inevitably subvert idealistic childhood expectations. Grant makes room for speculative options in the face of presumably objective truths by crafting story worlds that position impossibilities as undocumented anecdotes of historical events. This interstitial space between truth and speculation serves as the hypothetical workshop of *Into the Drowning Deep*, where Grant's novel fleshes out the mermaid of legend with seafaring idioms and granular scientific detail. Told as a revenge quest, it assumes mermaids are not just real, but deadly. Read through the lens of ecocriticism, the tale cautions against the dangers of neo-colonialism disguised as scientific discovery when researchers discover their subject supplants humanity's position as apex predator.

The story relates events of two scientific excursions to the Mariana Trench seeking proof mermaids exist; most of the action takes place aboard two ships named after mermaids, *The Atargatis* and *The Melusine*. By frequently juxtaposing discourses of science and legend, Grant emphasises the instability of boundaries between the two; she further accentuates such comparisons using recurring themes of paradoxical values that share surprising areas of overlap. Such examples include light and dark; science and entertainment; aesthetics and utility; seen and unseen; land and sea; and verbal and non-verbal communication.

The overlapping parts of these concepts echo environmental humanities scholar Stacy Alaimo's concept of *transcorporeality*, which "implies that we're literally enmeshed in the physical material world, so environmentalism cannot be an externalised and optional kind of pursuit, but is always present" (Kuznetski 2020). The novel highlights the permeable boundaries inherent to a sense of interconnected species across vast distances. As Alaimo notes, "there's no nature that we just act upon. Instead, it's also acting back upon us, as we are always already the very substance and the stuff of the word [*sic*] that we are changing" (Kuznetski 2020).

Adding to the ubiquitous liquid examples, the text's structure also mimics such fluid boundaries via paratextual divisions named for the ocean's ecological zones; starting at the surface, the titles mirror a downward progression through the Pelagic, Photic, Aphotic, Bathypelagic, Abyssopelagic and Demersal zones. Light penetrates only through the photic zone; similarly, as readers progress deeper into the text, the boundaries between categories become less visible. The metaphoric spatial inversion also corresponds to the characters' gradual descent into literal darkness and death. In addition to the section titles, Grant prefaces each chapter with fictive reports formatted as biographical excerpts, online forums and university lectures. The layered paratextual elements add weight to the intradiegetic assumption of mermaids' existence. As the plot moves forward, readers and characters face multiple zones of liminality with greater possibilities for interpretation.

As the sole funder of both expeditions, mass-media corporation Imagine Entertainment actively mediates most actions, thus further complicating the story beyond "human vs. nature." After rumours of senility fail to dethrone founder and CEO James Golden, the former "King of Schlock," he expands into television to enhance his legacy beyond bad sci-fi movies. The first expedition took place seven years ago during a planned "mockumentary" about mermaids, in alignment with the network's brand. As Grant notes, "[i]t seemed the public's thirst for cryptozoological fiction thinly veiled as fact was insatiable" (Grant 2017: 3). Nobody expected the ship to *find* mermaids until *The Atargatis* went radio silent; six weeks later, the US Navy found the ship abandoned.

While Imagine seeks to mediate narratives of both science and legend, leaked footage of the ill-fated voyage emerges online leaving them powerless to control increasing rumours. Grant's narrator discloses that "everyone [...]

in the corporation knew that *The Atargatis* hadn't been lost; she had been attacked. They knew that the video footage was real and unaltered" (Grant 2017: 66). The grainy film shows a horrific creature vaguely resembling a mermaid, "pulling itself along the *Atargatis* deck with clawed hands. Its tail was broad and flat [...] like an eel's [...] and while the substance growing from its scalp could have passed for hair in the right light, it was clearly something ... else" (55). Relayed in flashback and allusion, this deadly past informs present events. It also underlines the significance of water in the cycle of life and death.

The oceans' unseen depths and denizens motivate Alaimo's use of the phrase "blue humanities" to emphasise the "environmental orientation of oceanic and other aquatic scholarship [...] since most aquatic zones, species, and topics exist beyond human domains, requiring the mediation of science and technology" (Alaimo 2019: 429). Technological mediation is a central feature in Grant's novel as the crew sails to the deepest parts of the ocean: over 10,000 metres (6.8 miles) below the surface. Technology makes human research possible at such depths, as light fades significantly at 200 metres (660 feet) and reaches total darkness at 1,000 metres (3,300 feet); at depths of 4,000 metres (13,100 feet) scientific equipment endures extreme pressure – over 5,850 psi – that would crush an unprotected human (NWS 2017). In contrast to surface-level "green" ecologies, Alaimo describes these deep-water zones as a "violet-black ecology [...] one thousand metres down and much deeper [...] The deep seas epitomize how most ocean waters exist beyond state borders, legal protection, and cultural imaginaries" (2013: 233). The novel leans into this implicit sense of hiddenness and a fear of what lies below the surface to invoke Gothic elements with increasing visibility.

To determine the truth of the mermaid footage, Imagine eagerly funds a follow-up mission and fills it with scientists, storytellers and a complement of security "staff" from central casting. Included among them is Dr Jillian Toth, the world's first "sirenologist." She declined to join *The Atargatis* when, after her research, "she'd seen the shadow of a creature she hadn't wanted to meet" (Grant 2017: 73). Driven by guilt, Toth joins *The Melusine* feeling indebted to the crew who sailed "using information I provided. Because mermaids have [...] been luring us and drowning us for centuries. I want to see their faces [...] and know I was right every time I said they were out there" (105). Though reluctant to work with Imagine, Toth seeks validation for research that cast a Barnumesque shadow over her reputation among "serious" scientists.

Marine biologist Victoria Stewart (Tory) serves as the main focal character and eagerly accepts Imagine's invitation to serve as the ship's sonar specialist. Still mourning her sister Anne – a rising TV journalist until her assaignment aboard the *Atargatis* – Tory seeks answers in the song of the sea. She views the opportunity as a scientific windfall with the fringe benefit of potential revenge, oblivious to corporate machinations that would provoke drama by exploiting her grief. Also, among *The Melusine*'s crew are Tory's lab partner Luis Martines and three scientific siblings: identical twins Heather and Holly, both deaf, and older sister Hallie Wilson, who acts as translator and ASL specialist. Grant's details of scholarly pursuits paint the scientific community as proprietary, verging on the predatory, as she notes: "Scientists liked to argue about discoveries almost as much as they liked to make them, and arguing about someone *else's* discoveries was the best game of all" (44).

Imagine hires Olivia Sanderson as a journalist in Anne's former position. Despite crippling social anxiety, she creates physical and emotional distance to help mitigate her fears. Ray, Olivia's cameraman, friend and protector, owes his mobility to Imagine's medical research investments, as does corporate spokesman – and Dr Toth's estranged husband – Theo Blackwell. Just as the deep sea requires technology to facilitate human experience, each of these characters uses a mediating layer to bridge the gap between themselves and the real world. Tory and Luis experience the ocean through sound; the Wilson sisters communicate via sign language; Olivia projects a curated persona through the camera lens, and Ray and Theo rely on semibionic implants and medical technology for basic motor functions. As Grant's narrator explains, "Imagine was in entertainment. If it wanted to invest in loss leaders and philanthropic therapies [...] the world reaped the benefits, and Imagine made millions, while repairing a few cracks in its public image" (85). With the possible exception of Dr Toth, the entire crew remains metaphorically blind to awaiting dangers.

Naming is Owning

After the *Atargatis* tragedy, Golden seeks to bolster his company's image by naming the creatures at the possible expense of the crew of *The Melusine*. This neo-colonial intent overshadows any altruistic elements of the exploration giving it a far greater Gothic emphasis. David Punter points out colonising strategies at work in the Gothic when "any instance of naming in literature signals a new beginning or origin" (2014: 21). Such "new" origins ignore and attempt to overwrite, or colonise, existing histories; likewise Imagine want to recreate their history by owning the present.

Corporate executives front-load *The Melusine*'s expedition with non-disclosure agreements (NDAs), pre-arranged intellectual property rights, state-of-the-art scientific equipment and a ship outfitted with defensive shutters. Imagine spokesman Theo frames the voyage as one of discovery instead of revenge: "If we can find a mermaid, if we can prove these lovely ladies of the sea are more than just stories, we can answer a question humanity has been asking for millennia. We're not just here to right a wrong. We're here to make history" (Grant 2017: 97). Humans figure as the inherent problem in this quest for discovery. Alaimo explains the real-world scientific race against deep-sea mining that parallels the novel's quest:

> Since so little is known about so many species in the deep and pelagic seas – including, most intriguingly, the creatures that likely exist but have not yet been discovered – it is unlikely that there are not many species in the deep seas that have already become extinct due to anthropogenic causes. (Alaimo 2019: 430)

Laser-focused on possible discovery, the *Melusine*'s crew ignores the possibility of their potential role as prey.

Grant models her fictional mermaids on anglerfish, recognisable by the bioluminescent "lantern," or *esca*, that dangles above jaws crowded with needle-like teeth too numerous to contain (Monterey Bay Aquarium). As W. John Smith notes, these deep-sea predators "even mimic their prey's prey: angler fish [...] can wriggle fleshy outgrowths of their fins or tongues and attract small predatory fish close to their mouths" (Smith 2009: 381).

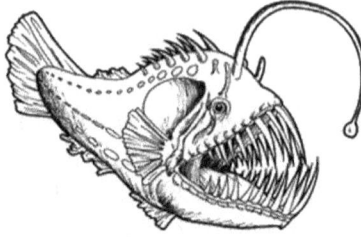

Figure 41. Female anglerfish (above) can outsize their male counterparts by several orders of magnitude (Image: Public domain).

While mimicry in literature describes survival strategies of oppressed peoples in post-colonial studies, Grant portrays mimicry as a form of audible bait in her text, just like female anglerfish use bioluminescent *esca* (see FishBase 2021). Bearing no resemblance to romanticised mermaid portraits, the creatures attacking them are not the "lovely ladies of the sea," but very clearly Other. As the novel's primary subject expert, Toth affirms their non-mammalian nature, but questions their biology:

> We can't be sure the mermaid is female [...] Everything about it suggests a human woman, but nothing about it is going to fulfill that promise [...] The only things out of place are those lips [...] Why would something so evolutionarily perfect need those lips? They sing in their throats. They speak with their hands. (Grant 2017: 241)

Similar to anglerfish, the fictional creatures exhibit extreme sexual dimorphism. Here, layers of detail accumulate to question boundaries of gender and sexuality. Grant focalises events through non-binary characters and reverses dominant gender roles; such reversal also alludes to the ancient Syrian goddess, Atargatis. Often called the first mermaid, Atargatis protected water, the universal other, and destiny. Fish were sacred symbols of life and fertility and fishing in her pools was forbidden. In later cults eunuch priests dressed as women and novices castrated themselves and performed tasks usually assigned to women. (Strong 1913: passim). In similar fashion, Alaimo confirms how nature breaks down gender dualisms, "as many species change their gender and sexuality, species and organisms change and emerge, making nature very fluid" (Kuznetski 2020).

Toth reveals their attackers as male, making the word *mermaids* inaccurate; she pronounces them *sirens* instead. "They learned to call for us because we wouldn't follow a shiny light" (Grant 2017: 243). The sirens illustrate the concept of Mertensian mimicry, which Smith describes in greater detail: "Many predators [...] fit unseen into their habitats. Others rely on the technique adopted by a wolf in sheep's clothing – they mimic a harmless species" (Smith 2009: 381). Dawning realisation about the unseen female's prodigious size and the chaotic consequences of her surfacing forces the rising action to resolution. The "harmless" humans manage to capture a live specimen while sirens swarm the ship. Eventually, unlike their predecessors, some survivors disembark *The Melusine*.

Conclusion

While myth portrays mermaids as alluring human hybrids, Grant's sirens present a terrifying image of the deep sea, yet Toth describes them as "perfect predators" that need protection. Conservation applies to all members of an ecological system, not just the pretty ones: survival depends on biological diversity. As Toth points out, the sirens must be protected, "because it's our job. We don't just conserve the things we like, or the things we find adorable" (Grant 2017: 329). In contrast, Imagine hires male security guards based on their physically attractive qualities; they present an illusion of security, yet make better bait than bouncers. Frequent comments about the staff as useless eye candy further emphasises the issue of aesthetics vs. utility, but when rescue teams arrive, they transport every survivor back to shore, not just the prettiest.

Alaimo probes the value of aesthetics when she points to spectacular representations of deep-sea creatures, portrayed "as astonishingly weird or beautiful" (Kuznetski 2020). The mermaids, sirens and security staff all represent stereotypes of groups that include more than one essential type. Images of the deep sea circulating in popular culture tend to be sensationalised or stylised in some other fashion. Alaimo further notes "[t]he often highly aestheticized aquatic figures underscore how the arts and humanities are essential for understanding the scientific captures and dissemination of particular narratives, tropes, and styles across science, literature, art, popular culture, and activism" (Alaimo 2019: 431). In other words, science needs the humanities to tell stories that contextualise images and discoveries; stories help us develop empathy while providing explanations in human terms. As Alaimo emphasises, "[a]rt, literature and popular culture can make scientific facts and data into something much more meaningful for people" (Kuznetski 2020). Folk and fairy tales like *Melusine, Undine*, or *The Little Mermaid* may not paint a realistic picture of deep-sea creatures, but they do communicate a human need for connection with Others. "More generally," Alaimo reflects "I think that we need the human imagination to enliven and contextualize scientific that discloses otherwise invisible processes and effects" (Kuznetski 2020).

Just as the crew aboard the *Atargatis* fell prey to the same elemental forces protected by the eponymous goddess, the crew of the *Melusine* follows the fairy tale's path. A half-fairy, Melusine's curse changes her into a half-woman, half-fish (or serpent) every sixth day. In similar stories like the *Knight of the Swan* or *Psyche and Cupid*, Melusine marries with one caveat: her spouse must swear to honour her privacy in the bath. He inevitably discovers her hybrid form and betrays her by making the knowledge public (see Ashliman 2015 for variations). Crew members on the second mission discover sirens, but may not reveal their knowledge publicly, according to their NDAs. Jillian Toth and Theo Blackwell also reflect versions of the tale in which a water spirit marries a mortal man. So much understanding depends on how we choose to use the tools available, such as the stories we choose to tell, how we tell them and whether or not we omit some details, either as a deliberate choice or benign neglect. Similarly, as more obvious tools, scientific instruments can quantify or record data, or become weaponised by the less scrupulous.

Stories that provoke emotional reactions – whether anger, horror, or humour – have the power to inspire readers, elicit questions and raise awareness about endless topics. Fiction that operates outside the canon in liminal zones like sci-fi, horror and fantasy has the added benefit of making the impossible seem plausible, if not probable. Such zones of liminality, whether ocean zones that overlap and obscure one another or various fictional genres, prompt us to ask more interesting, productive questions about who we are and how to be in the world. If not for the haze of overlapping boundaries, we may not seek clarity for ourselves.

Neo-colonialism and the Liminal

Catherine Pugh

Crawl (Alexandre Aja, 2019)

The 2019 film *Crawl* follows a young woman, Haley (Kaya Scodelario), her estranged father, Dave (Barry Pepper) and their dog, Sugar (Cso-Cso) as they are hunted by alligators while trapped in the crawl space underneath their house during a Category 5 hurricane in Coral Lake, Florida. Throughout the film, eco-horror aspects work to destabilise and transgress boundaries of what is considered to be human, and what is considered to be home. Although animals are regularly positioned outside of human ethics, they are continually anthropomorphised, which in horror texts endows them with human desires and motives to send them down a path of violence and vengeance. Animals in horror are antagonists either because they are simply violent and therefore their behaviour cannot be helped, or because they desire to kill and maim. Either way, they are branded the Other: the opposite of the supposedly civilised, cultured and ethical human.

Initially, the alligators of *Crawl* appear to be the primary aggressors, but closer reading reveals them to be agents of the bigger threat – a savage and malicious Nature that launches an assault on a landscape it once ruled. By destroying the protagonists' home – a symbol of their family as well as their implied place in the food chain – the attack paradoxically manages to reunite the estranged family, while allowing Haley to emerge as the "apex predator" her father always insisted she was.

Ecological Decolonisation

Animal Horror, Eco-Horror and the Home

Although *Crawl* is ostensibly an animal or Nature horror film, it can also be categorised as an eco-horror text. Eco-horror has been referred to as "Nature Run Amok" (Dan Whitehead 2012) or "Revolt of Nature" films (Kim Newman [1988] 2001: 91), where Nature itself becomes a monstrous threat to humanity. Katarina Gregersdotter, Nicklas Hållén and Johan Höglund (2015a: 32) argue that these creatures do not necessarily seek revenge, but are instead attempting to "reclaim their central position in the ecosystem and on Earth."

In eco-horror, threats to the protagonist can come from any aspect of Nature, such as the weather, plant-life and land/seascape as well as animals. Although eco-horror often plays a part in animal horror, particularly in texts where animals react aggressively in response to humankind's intrusion or violation of nature, Gregersdotter, Hållén and Höglund (2015: 4) argue that these films equally "centre on the relation between 'human' and 'animal' categories unrelated to their places in the eco-system." *Crawl* features both animal horror and eco-horror elements; while the might of Nature is made monstrous through weather, flooding and animal attacks; the alligators themselves are frequently referred to as Haley's counterpart in through both imagery and dialogue. Notably, throughout childhood flashbacks Dave refers to Haley as an "apex predator" to psyche her up for swimming competitions. During the attack, she rejects this description of herself, only to reclaim it later when she is able to out-swim the alligators. Paradoxically, then, Haley must indulge in her primitive "animal instincts" in order to preserve her place on the food chain.

Furthermore, unlike many eco-revenge narratives, Haley and her father have not (accidentally or otherwise) intruded into the alligator's territory; they are not being punished for transgressions against nature. They and their neighbours have lived peaceably alongside the alligators in Coral Lake for many years and there are no plans to build on or drain the swamp that would catalyse the revenge of Nature. Instead, Nature attacks because it can, to reclaim its "home"; a motive reinforced by the film's tagline "They Were Here First." By doing so, it forces Haley and Dave to re-evaluate their own ideas of home. As the film begins, the two are estranged, primarily communicating through

Haley's sister. Haley's mother and Dave have recently divorced, with the mother now abroad with her new partner. Haley and Dave's relationship is strained, with Haley believing that Dave neglected his marriage after spending all of his time working as her swimming coach. Haley goes looking for her father during the storm, eventually discovering him at the old family home, which is up for sale. As the house gradually floods, Haley finds him injured in the crawlspace under the building where the water is coming in. Floor by floor, the house is erased; its status as a safe hold, a boundary construct and symbol of the united family is washed away. Furthermore, both human occupants become caught within its trappings: Dave almost drowning in the crawlspace, while Haley is forced to traverse the overflow pipe in order to escape. Here, the house is not only obsolete, but also dangerous.

However, the main structure of the house – and therefore the family – ultimately survives; it is "a strong one" as Haley puts it. Towards the end of the film, Dave admits that he could not bear to sell the house as "It's our home. The last place we were a family," to which Haley replies "This house isn't our home. Me and you, that's home." Dave is afraid that letting go of the house also means letting go of the family; while the film returns the landscape to Nature (at least temporarily), it nevertheless affirms that this reclamation does not destroy family bonds. Nature (the biological phenomena) is not as strong as nature (the essence of what holds a species together). The alligators also make

Figure 42. The domestic is erased by the wild in *Crawl*, directed by Alexandre Aja (Paramount Pictures, 2019).

the house their home, laying eggs in a subterranean, vegetation filled environment; ancient creatures slowly returning to (and, in turn, reforming) their space. The film appears to insinuate that humans need to vacate these spaces to make room for the animals that previously inhabited the environment, albeit in a macabre and blood-thirsty way. Despite Haley and Dave injuring and trapping the alligators, they rarely kill them, only doing so when they are in mortal danger (unlike the vast majority of animal horror films, where the animals are seen as mindless villains to be destroyed at all costs). There is an underlying begrudging respect between humans and animals, reinforced by the parallels drawn between Haley and the alligators throughout the film.

The alligators kill indiscriminately and are therefore not actively targeting Haley and Dave, rather it is Nature itself that is the primary antagonist, one that seems to act specifically against the father and daughter. As stated in numerous radio broadcasts during the film, humans – and human structures such as buildings and cars – are under threat from the extremely powerful hurricane even before the alligators arrive, therefore it can be argued that Nature itself

Figure 43. Haley in the alligator nest: the outside brought inside in *Crawl*, directed by Alexandre Aja (Paramount Pictures, 2019).

actively and viciously pursues Dave and Haley. Towards the end of the film, Dave, Haley and Sugar have escaped the house and begin to make their way to a nearby boat. Dave believes they can reach it as long as they move slowly as the alligators "only hunt what splashes" and the heavy rain should camouflage them. However, at that moment the eye of the hurricane passes over them, turning everything still and quiet. Suddenly left with no rain to cover their movements, the group becomes stuck as the alligators silently glide around them. Fortunately, Haley is able to out-swim the alligators and return with the boat, but as she retrieves Dave and Sugar, the levees break and the giant wave of water washes all of them back into the now-flooded house. It becomes clear that Nature is the threat in *Crawl*, the alligators are simply part of her armoury.

This idea is underlined by the first appearance of an alligator in the film as Haley drags her injured father through the crawlspace. Unlike other animal horror texts, such as *Jaws*, there are no predatory point-of-view (POV) shots of the alligator stalking Haley, no musical cue, tension-building cuts or focus on dead spaces to indicate that something is watching from the shadows. The

Figure 44. Dave, Sugar and Haley attempt to cross the water without drawing attention in *Crawl*, directed by Alexandre Aja (Paramount Pictures, 2019).

alligator's sudden and violent arrival seemingly comes out of nowhere, echoing a tree that earlier crashed through the kitchen window and frightened Haley. The tree foreshadows the alligators' arrival, just as towards the end of the film another alligator bursts through an upstairs window as Haley stares out. Nature attacks the house and the people inside, disrupting the perceived sanctuary of the home by (almost literally) throwing its agents against the walls with such force that it breaks them down.

Destabilising Anthropomorphism and Empathy

The alligator's arrival is typical of the unusual lack of anthropomorphism in *Crawl* compared to other films of its kind, which tend to endow the animal with human desires such as revenge. Paradoxically, animals in these texts are also seen as naturally savage, forcing the human to regress to a devolved state in order to survive. Gregersdotter, Hållén and Höglund (2015: 7) write that, "The animal is hardwired to be a relentless predator, unable to show remorse or pity. Therefore, the only way for humans to protect themselves against the ferociousness of the animal is to respond to it by becoming as ferocious as the animal, and to kill it." Haley is consistently told to be more like the alligators, to out-swim and outsmart them because she is the "apex predator." Even in the opening scene where Haley is taking part in a swimming competition (which she narrowly loses), a flashback to her childhood shows her father making her repeat these words, foreshadowing how essential this phrase will become to her identity. Later in the film, when she is pitched against the alligators, the competition is renewed, albeit with much higher stakes. This time, she manages to beat all her "competitors" to reach the boat, triumphantly declaring "Apex predator all day!"

A typical technique used in animal horror (or any horror that involves stalking) is the POV shot, allowing the spectator to see through the eyes of the "killer," to build tension and to demonstrate the animal's consciousness. *Crawl*, however, rarely employs this technique, interrupting potential anthropomorphism or empathy. Instead, overhead shots are used to great effect, giving a bird's eye view of the events. The attacks are bloody and are over relatively quickly. The focus instead shifts to humans seeing the alligators coming towards them, with the main feature being the inevitability and sheer violent speed of these attacks rather than a slow build-up or predatory gaze of the animal. Furthermore, the number of creatures involved degrades anthropomorphism: the alligators have no distinguishing features or names, and when one dies there is always another to take its place, more akin to zombies than other alligator and crocodile horror films such as *Alligator* (Teague, 1980), *Dark Age* (Nicholson, 1987), *Lake Placid* (Minor, 1999); *Black Water* (Nerlich and Traucki, 2007), *Rogue* (McLean, 2007), *Primeval* (Katleman, 2007) and *Black Water: Abyss* (Traucki, 2020), all of which feature an individual or primary animal antagonist.

Another notable cinematic technique that anthropomorphises the animal involves a close-up of the animal's eye as it watches the human, creating a connection between the two. As Gregersdotter and Hållén explain, "close ups of the animal eye are a reminder of the simultaneous closeness and remoteness of the world of the animal's psyche" (Gregersdotter and Hållén 2015: 219). However, *Crawl* lacks these eye-to-eye shots. Haley never meets the gaze of any of the alligators and they do not consciously search for her. Rather, their poor eyesight means that they attack whatever they can actually see. The lack of

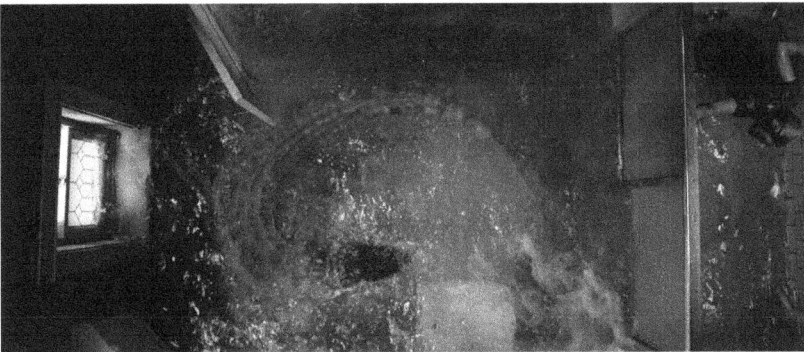

Figure 45. An overhead shot of the bathroom as one of the alligators prepares to attack Haley (Kaya Scodelario) in the shower stall on the right. *Crawl*, directed by Alexandre Aja (Paramount Pictures, 2019).

stalking in *Crawl* underlines the disconnect between animal and humankind; the intimate relationship between monstrous predator and prey is corrupted, simultaneously disrupting structures of difference. Despite the comparisons between Haley and the alligators, the latter are very definitely Othered. Instead of acting as a bridge between the two species, the eye is actively destroyed. Dave points out that the alligators cannot see very well, banging on metal pipes to disorientate them, while Haley blinds at least two of the creatures. Rather than acting as a counterpoint to humankind, the alligators appear as part of the threatening landscape, brought in and disguised by the water.

Incidentally, Sugar the dog plays an important part of tempering the relationship between the human and non-human animal in this text. Sugar is loved and cared for, acting as proof that the humans here are neither malevolent nor deserving of punishment. Equally, Sugar's presence supports the idea that if Nature has an agency in this film, then her viciousness is aimed at humans only, as never specifically attack the dog even when she is in the water alone.

The animals of eco-horror may be at the opposite end of the spectrum from the sentimental benevolent animals in other films, but both have a tendency to bring people together, particularly families (see Burt 2002: 115). The horror animal retains the function to heal and reunite, with aggressive attacks indirectly reconciling damaged relationships by threatening the family unit. The devastating animal attacks in *Crawl* nevertheless renew the estranged family's bond as well as forcing them to talk about the issues that pushed them apart. The animal horror film perverts the sentimental trope of the family learning to heal through the wise and kind animal; instead, it is by standing against Nature – or running away from it together – that unifies the family. Jonathan Burt (2002: 187) writes that, "[d]eath or divorce in the family is, after all, a threat to family cohesion, rupturing the 'natural' ties of child and parent, just as natures is threatened by various forms of development". However, instead of tales "outlining how one should behave and the manner in which animal and human relations are supposed to mirror each other," animal attack texts offer a monstrous threat that underlines the violent and aggressive side of human nature. Burt (2002: 187) also notes that the healing animal has a particularly powerful effect on traumatised or unhappy children, who become transformed through their relationship with the animal. While Haley is indeed transformed, becoming a more confident, determined "apex predator", this is by aligning herself with the wild threat and then effectively being "better at it" than them. Her father constantly reminds her that she is "faster than them" and she ultimately proves herself just as cunning and vicious; outswimming, outsmarting and outfighting the creatures, even able to remain calm enough to escape when one has her clamped in its jaws in a death roll. Animal horror demands that the the survivor is better at being a beast than the creatures themselves. Rather than learning empathy, as is the want of other animal or power-of-nature texts, protagonists must effectively beat them at their own game by being "wilder", by placing themselves outside of ethics.

Consuming Eco-Horror

Animal horror primarily relies on the threat of being consumed, usually by being eaten (see Adams 1990; Gregersdotter, Hållén and Höglund 2015a; Piatti-Farnell 2017). The alligators of *Crawl* do not appear to attack for food; although they rip bodies apart, they are not seen swallowing these body parts, even leaving a body almost intact by the nest (presumably to be eaten by the emerging infants). Their threat lies in their violence – their "wildness" – not their hunger. However, a nod to this fear is made when a doomed looter at a nearby petrol station eats a hotdog shortly before he is attacked by one of the alligators. The sudden reversal of consumer to consumed is a violent assault on humankind's position as the apex predator, further destabilising the already contentious separation between human and animal.

In this sense, eco-horror can be considered a subgenre of what Carol J. Clover termed "urbanoia" (where urbanites become victim to the savage countryside and its inhabitants), or even a type of rape-revenge story with "Mother Nature" cast as the final girl. Clover describes the typical villain of urbanoia as "beyond the reaches of social law" (Clover 1992: 125), much like Gregersdotter, Hållén and Höglund assert the horror animal as beyond the ethics. Clover's description of the urbanoia villain is reminiscent of the threat of animal horror, with emphasis on poor manner and hygiene, particularly teeth (Clover 1992: 125–6). The immediate threat of the animal is in its killing potential; in its teeth and claws and supposedly insatiable appetite. While Clover argues that it is "economic guilt" that drives city-revenge or urbanoia films as "city comforts are costing country people their ancestral home" (1992: 133), the parallel world of eco-horror turns on both economic and environmental guilt. Unlike urbanoia films, however, the animal attacks of eco-horror can be a just, if extreme, punishment for human cruelty and transgressions. Rather than the wilderness attacking because it is beyond ethics, it attacks for justice and/or vengeance, more akin to the rape-revenge horror film.

In *Crawl*, the modern, suburban family is pitted against ancient, swamp-dwelling creatures. Left without weapons for the vast majority of the film, Haley and Dave are forced into the role of prey, relying on their knowledge of the natural world to help them survive, such as banging on pipes to disorientate the alligators so that Haley can slip past them. Like the urbanoia film, the protagonists are left helpless, without technology, defences, or assistance, at the mercy of the threatening wild and its inhabitants that are both beyond ethics.

Animal horror thrives on challenging the separation of animal and human. Even narratives that highlight the differences between the two assign them human desires and qualities. *Crawl* demands that Haley recognise her potential as an apex predator and therefore a survivor. Destabilising the boundaries between human and non-human animal, house and home, outside and inside, allows a structural collapse that ultimately allows the humans to relinquish the prized home to Nature. Nature in *Crawl* is neither a healing nor avenging force; it is a testament to the violent potential of unchecked Nature and its power to devastate, and reclaim, everything in its path.

Tom Ue

Spirited Away (Hayao Miyazaki, 2001)

Hayao Miyazaki's Academy Award-winning animé *Spirited Away* (2001) has inspired a devout following amongst viewers internationally.[1] A recent survey of 177 critics ranked it fourth amongst the 100 greatest films made in the twenty-first century; and at the time of writing, it places 31 in the top-rated movies on IMDb. *Spirited Away* opens with the 10-year-old Chihiro (Rumi Hiiragi) and her parents (Takashi Naitô and Yasuko Sawaguchi) driving to their new home. Chihiro's father takes a wrong turn. When her mother gestures to their new house, the last of a row perched on top of a hill, he admits, "I must have missed the turn off" (195). Rather than reversing their vehicle, however, he presses on, brushing aside his family's reservations and relying on his intuition and their Audi four-wheeler: "This road should get us there. [...] Trust me. It's gonna work" (195). He ushers his family into the unknown until they arrive at an old building. The family leaves their car, discovers that this structure is rather new and made of plaster, and follows its tunnel into an abandoned theme park that doubles up as a spirit world. All the while, Chihiro's parents ignore her objections. At one of many empty restaurants there, they greedily devour the food on platters, indifferent to the fact that it is meant for spirits. Chihiro's

1   I thank Simon Bacon and Marko Teodorski for many kinds of help; and Noel Brown, Mary Beth MacIsaac, and Gareth Reeves for our conversations about *Spirited Away*. I am grateful to Dalhousie University Libraries and the Social Sciences and Humanities Research Council of Canada for supporting my research. References to *Spirited Away* are taken from *The Art of Spirited Away* (VIZ Media LLC, 2013): the story and screenplay are by Hayao Miyazaki, and the English-language adaptation is by Cindy Davis Hewitt and Donald H. Hewitt. My revisions to the quotes are entered in squared brackets.

parents are transformed into pigs for their transgressions. *Spirited Away* traces her quest to reverse the spell and to remove the family from this enchanted world. In this chapter, I advance scholarship on Miyazaki's eco-commentary by directing our attention to his treatment of exchanges. It is no coincidence, I argue, that the film's very first shot is of a card and a farewell bouquet that Chihiro received from her best friend Rumi. If, on the one hand, they remind us that endings can also be interpreted as be-ginnings, then on the other, they accentuate how the family is trading their old life for a new one. Their new home may be, as Chihiro's mother notes, "in the middle of nowhere", but it is one that, according to her father, they "[wi]ll just have to learn to like" (195).

This essay reveals how Miyazaki uses exchanges to meditate on our re-lationship with the environment. The film's central setting, Yubaba's (Mari Natsuki) bathhouse, is a case in point: it is profitable – the witch is regularly counting money and she lavishes gifts on her Baby (Ryunosuke Kamiki) – but it operates at significant expense to the natural world. This establishment burns coal and, as Dani Cavallaro (2006) observes, its "chimneys discharg[e] black smoke into the surrounding blue sky" (141). The theme park, of which the bathhouse operates as a kind of synecdoche, is one of many built in the early 1990s that went bankrupt when the economy went bust. Standing on a bank of rocks, Chihiro's father observes: "Look. They were planning to put a river here" (197). As Nathalie op de Beeck (2009) remarks, "[t]o build the park, ground was broken and a river rerouted. Now the useless town has fallen into complete disrepair, but the past wilderness cannot come back as quickly as it was destroyed" (275). Miyazaki is responding, as Susan Napier (2018) writes, to a Japan "mired in a culture of materialism that seemed an effort to allay a debilitating sense of spiritual emptiness through incessant consumption:

> The government, in an attempt to jump-start the nation's long economic decline, had for decades promoted huge construction projects around the country. The flood of concrete destroyed forests, poured over beaches and mountains, and dammed up virtually every waterway that still flowed freely. Fast-food outlets and vending machines invaded the highways, and convenience stores took over city streets. (200)

In the film, Chihiro is quickly immersed into the spirit world and exposed to its economic innerworkings. Haku (Miyu Irino) instructs her to make herself useful to save her parents. She must gain employment from Kamaji (Bunta Sugawara), the boilerman: "Tell him you want to work here, even if he refuses, you must insist. [...] If you don't get a job, Yubaba will turn you into an animal. [...] Kamaji will try to turn you away to trick you into leaving, but just keep asking for work" (201–2). The overworked and many-armed Kamaji is "slave to the boiler that heats the baths" (203) and he is ably assisted by his many sootball workers. "I don't need any help", he asserts. "The place is full of soot. I just cast a spell on them and I've got all the workers I need" (203). Kamaji warns the sootballs against indolence. As he explains to Chihiro, "[i]f they don't work, the spell wears off. They turn back into soot" (203). He nevertheless helps her by bribing, with a roasted newt, Lin (Yumi Tamai) to take Chihiro to Yubaba. He tells Chihiro: "If you want a job, you'll have to make a deal with Yubaba. (pause) She's the head honcho here" (204). Haku's and Kamaji's warnings and directions make pronounced how so many of the characters, from Yubaba's apprentice Haku down to the sootballs, labour in a capitalistic economy.[2] Redundancy makes them expendable. Yubaba would happily repurpose Chihiro into a piglet or a lump of coal – commodities that her business can use – but she decides, instead, that Chihiro would be more valuable otherwise. "I'll give you the most difficult job I've got", she says, "and work you until you breathe your very last breath" (206). Yubaba rushes Chihiro into signing a contract, one that grants her little by way of protection (see Figure 46). The witch wouldn't hesitate, as she routinely warns, to transform Chihiro into a pig or coal if she complains. When she praises Chihiro, as she does later in the film, it is in terms of the money that she brings in: "Sen, you did great! We made so much money! (pause) That spirit is rich and powerful. Everyone, learn from Sen" (217).

Yet Miyazaki reminds us that we are more than our jobs by exposing us to alternative kinds of economies. Haku, Kamaji, and Lin all risk Yubaba's wrath to help Chihiro. They offer, and are recompensed with, friendship; and they care for each other in ways that are unknown and unknowable to the witch. Miyazaki's concentrated energy on exchanges furnishes us with a vocabulary

2   In this, Yubaba recalls Chihiro's parents, who believe that money can purchase anything and everything. At the restaurant, Chihiro's mother is certain that they "can pay the bill when [its proprietors] get", while her father reassures her: "Don't worry, you've got Daddy here. He's got credit cards and cash" (198–9).

Figure 46. Yubaba rushes Chihiro into signing a contract. *Spirited Away*, directed by Hayao Miyazaki (Walt Disney Home Entertainment, 2001).

for thinking about our relationship to the environment, and we'd profitably read *Spirited Away* alongside Margaret Atwood's *Payback: Debt and the Shadow Side of Wealth* (2008). Regarding Charles Dickens' interlaying of economic and moral debts in "A Christmas Carol" (1843), Atwood claims: "By being a creditor of such magnitude in the financial sense, [Scrooge] himself has become a debtor in the moral sense, and it's this realization that's at the core of his transformation" (171). She goes on to adapt, and to give a new line of application to, Dickens, reconfiguring Scrooge to point to what we owe the environment. Atwood's "Scrooge Nouveau" is sleeker than Dickens', and he more readily lavishes on himself. He is also much wealthier. Scrooge Nouveau collects corporations, and he is indifferent to "what they make, so long as they make money" (175). The Spirits of Earth Day Past, Present, and Future show him the consequences of man's neglect and exploitation of the natural world. The final Spirit grants Scrooge Nouveau access to a possible future, one rife with suffering for one and all:

> Mankind made a Faustian bargain as soon as he invented his first technologies [...]. The end result of a totally efficient technological exploitation of Nature would be a lifeless desert: all natural capital would be exhausted, having been devoured by the mills of

production, and the resulting debt to Nature would be infinite. But long before then, payback time will come for Mankind. (201–2)

Atwood encourages us both to rethink our priorities and to reinvest in our world. "Maybe", she hazards, "we need to calculate the real costs of how we've been living, and of the natural resources we've been taking out of the biosphere. Is this likely to happen? Like the Spirit of Earth Day Future's, my best offer is Maybe" (203).

Spirited Away is rife with examples of man's abuse of nature, chief amongst them the pollution to and the damming up of rivers. I have mentioned the river in the theme park, but Chihiro's encounters with two River Spirits reveal the extent in which man has taken from these bodies without ever giving back. An early scene economically reveals how capitalism gets in the way of eco-efforts. Chihiro and Lin are tasked with scrubbing the big tub, a bath that hasn't been cleaned in a long time, that is reserved for the dirtiest customers, and that shares, with water bodies, the capacity to hold water. When capitalism – in the form of customers – beckons, Chihiro and Lin give up and they start up an herbal bath. As Lin explains, this steamy mixture effectively obscures how dirty the tub is: "Dried worm salts. It's supposed to be good for you. [...] And with water this murky, you can't see all the sludge in the tub" (213). The bathhouse prioritises money over its clients' comfort and well-being. Capitalism puts pressure on the Chihiros and Lins to conceal, rather than to remove, the grime; and its accumulation makes it now difficult, perhaps even impossible, to clean. The crucial project of improvement, Miyazaki seems to suggest, demands that we acknowledge and that we tackle its challenges head on. The tub's first guest is a River Spirit (Koba Hayashi), who initially appears, even to a practitioner of magic as experienced as Yubaba, to be a Stink Spirit. The Spirit would not be denied entry, and his odour overcomes all he passes. Yubaba, though suspicious, cannot determine what it is: "Something's fishy. That doesn't seem like a Stink Spirit to me ... But we have no choice. Go greet him. [...] Just give him a bath and get him out of here as fast as you can" (114). Seemingly unable to recognise itself and visibly in discomfort, this Spirit struggles to articulate to Chihiro his desire, firstly, for the tub to be refilled, and secondly, for the removal of all of the garbage deposited in its river (see Figure 47). This water body has variously served as a dumping ground, a food source, and a recreational spot. The Spirit breathes a sigh of relief when

Figure 47. A substantial amount of garbage, from bicycles to toilets, has been deposited in the river. *Spirited Away*, directed by Hayao Miyazaki (Walt Disney Home Entertainment, 2001).

Chihiko finally "pulls out the final plug" (117) – in the form of some fishing lines and bobber (Figure 48). These fishing gears hurt the Spirit as much as the bicycle handle lodged in its flesh.

In exchange for her help, the River Spirit rewards Chihiro with an emetic dumpling, half of which she will feed to Haku, the film's other River Spirit. Chihiro's and Haku's fates, it transpires, are intertwined. He tells Chihiro, enigmatically, in an early meeting, "I've known you since you were very small" (202) – offering no explanation for how he knows her without her knowing him. Chihiro will complete this puzzle when she comes to realise that he is a river. "Once when I was little", she says, "I dropped my shoe into a river. [...] When I tried to get it back, I fell in, thought I'd drown, but the water carried me to shore. [...] It finally came back to me ... (pause) That river's name was ... the Kohaku River. [...] I think that was you, and your real name is Kohaku River" (234–5). Chihiro's recollection and her restoration of Haku's identity reverse Yubaba's enchantment, and he transforms back to human. As op de Beeck (2009) persuasively argues:

Figure 48. Chihiro removes fishing gear from the River Spirit. *Spirited Away*, directed Hayao Miyazaki (Walt Disney Home Entertainment, 2001).

> Chihiro's sudden revelation, a strong memory of a simultaneously dangerous and benevolent river, suggests that a capacity to appreciate wildness might be the first step toward saving the land that is left. She redeems Haku, the embodied remnant of a river, with her commemorative words and with her stubborn action – but to do so, she had to become a warrior and lose her fear of the animistic, undeveloped world that she displayed at the film's beginning. (279)

Chihiro and Haku complete their memory together.[3] She explains how Haku's river was filled up and how it is now occupied by apartments. "That must be why I can't find my way home", Haku realises, "I remember you falling into my river, and I remember your little pink shoe" (235). Chihiro completes their story, emphasising her debt to him: "So you were the one who carried me back into shallow water. You saved me. (pause) I knew you were good" (235). Haku's actions in the past enable Chihiro to help him in the present. This exchange, while important for the characters, does little

---

3   Ghibli films regularly thematise time and memory. For a discussion of this phenomenon, see my essay "Narrative, Time and Memory in Studio Ghibli Films" (2014).

to unsettle the environmental problems that Miyazaki raises. Urban development has unmoored Haku, rendering him lost and homeless. He regains a kind of home in our memories, and perhaps Chihiro's too, but, surely, Miyazaki suggests, it's a poor substitute for the real thing.[4]

4   For an account of the changes made to the film in its American release, see Michael McCarrick's "The American Version of *Spirited Away* Makes a MAJOR Change to the Ending." According to Miyazaki, Chihiro does not, but she will eventually, remember, what happened in the spirit world. By the end of the the American version, Chihiro gains confidence from all of her adventures, and she reassures her parents that she isn't intimidated by "[a] new home and a new school": "I think I can handle it" (237).

Image Intervention III: Becoming Merfolk

Figure 49. Artwork by Gemma Files (Reproduced with permission).

Part VI

Ecological Entanglements and Environmental Futures

Tom Shapira

Abe Sapien (1994–present)

From its inception the comic book series *Hellboy* has shown great interest in the subject of the world after the end. Throughout the series both individuals and factions believed that the end of the world is coming. Rather than try to avert the apocalypse they dedicate much of their efforts to what comes later: "This is Ragna Rok – not only the end, but a new beginning" (Mignola 2001). The apocalypse is seen as something that cannot be averted, only mitigated. As the fictional universe progressed a considerable part of that effort was dedicated to the character Abraham "Abe" Sapien.

Originally a mere supporting character in the narrative, Abe appears as an amphibious fishman of unknown origin (see Figure 50). He is just another supernatural agent of Bureau of Paranormal Research and Defense (BPRD), a group that serves as the investigative arm of the United States government in all manners relating to the supernatural. As the *Hellboy* series gained more popularity it was given several spin-offs, including *BPRD* (2002), a team-based title in which Abe served a major role, and later his own personal *Abe Sapien* series (2013). In both we gain deeper understanding of Abe's history and his role in the future. When Hellboy seemingly perishes in his own ongoing title Abe takes his role as a figure the various factions struggle over, all under the belief that he is a character of great importance. His amphibious status becomes not just his ability to survive both in and out of water but also of the symbolic notion that he is a being of two worlds – the current one and the world to come.

In the miniseries *B.P.R.D.: The Garden of Souls* (2008) Abe is summoned to a small Island near Indonesia, in which he discovers a group of long-lived scientists. Kept alive by special mechanical bodies, these people were his partners in a previous, human, life. Before he became Abe he was Langdon Everett

Figure 50. *Hellboy: Seed of Destruction* #2 (Mike Mignola, 1994). Copyright Dark Horse Comics reproduced under Fair Use legislation.

Caul, a nineteenth-century explorer who came upon a mystical object that triggered his transformation. The group, the Oannes Society, believes the apocalypse is coming. Their plan to "save" humanity is to detonate a series of large bombs that would kill millions of people whose souls they will harvest into a few artificial bodies. The members of the Oannes Society will take over these new bodies and keep the spirit of humanity alive while everyone else perishes: "Yes, millions would lose their lives, but at a specific time of our choosing. Thusly, at that moment we can control their final fates" (Mignola and Arcudi 2007).

Members of the group are certain Abe would side with them, but he is horrified and rebels against the plan, rationalising that the salvation offered by the group is no salvation at all. Not only are the physical bodies of people destroyed, but the souls that are harvested would remain under the control of the Oannes Society. Michael E. Zimmerman would probably recognise the Oannes Society type of philosophy as Ecofascism: "An ecofascist movement would have to urge that society be reorganized in terms of an authoritarian, collectivist leadership principle based on masculinist-martial values" (Zimmerman 1995: 209). The Oannes Society fits neatly into this categorisation. Not only is it an all-male group (older white men specifically), but they also have little trouble in sacrificing individual lives, even in great number, in order to preserve a utopian vision of the land.

Even their choice to commit a mass murder near Indonesia (instead of the West) evokes ethno-chauvinism. As noted by researchers of over-population discourse, Western intervention in matters of ecology is often an excuse to control foreign bodies, who are seen as potential threat: "White men, attempting to control the reproductive capacities of women in the Global South, through family-planning programs with questionable intentions, is an example of this" (Dyett and Cassidy 2019: 210). With the Oannes Society "control" is quite literal and absolute. They care about "humanity" as a nebulous ideal: "Insofar as deep ecology fails adequately to recognize that human life has more values than other life forms, he argues, it promotes 'Ecofascism,' namely the sacrifice of individuals for the benefit of the ecological whole" (Taylor and Zimmerman 2008: 458). The millions set to die by the massive bombs becoming a sacrifice in some twisted religious ceremony.

While members of the Oannes Society consider themselves scientists first and foremost, their ideology is backed by religious zeal. Seeing Abe, one of their former comrades, transformed into a fish-creature lights the fires of fervour in their hearts.[1] The Oannes Society sees his transformation as sign of the righteousness of their cause. Their desire is not simply to save humanity, but to become the guiding light of a new civilisation. The many that will die during the catastrophe are not a side effect; they are a necessary sacrifice in the name of the new world order. Without the great catastrophe their vision for the world to come could not be complete.

---

[1]   That he goes against their plan does not deter them one bit, their religious zeal has a way to work around the "betrayal" of their prophet figure.

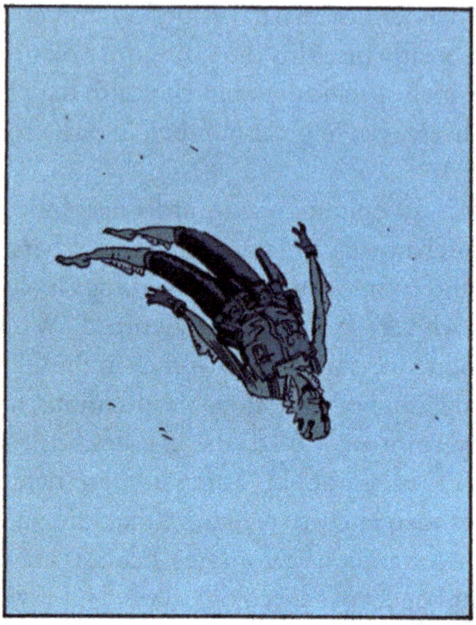

Figure 51. Image of Abe from *B.P.R.D.: Plague of Frogs* #5 (Mike Mignola, 2004). Copyright Dark Horse Comics reproduced under Fair Use legislation.

New Breed of Men

Abe's fishlike nature can be understood as having a relation to the primordial origin of life on Earth (see Figure 51).

Ecofacism's ties to Nazi ideology means that those who survive are not picked at random. They are the chosen, and that choice is often constructed in racial terms. Eric Kurlander, in his study of the supernatural in the Third Reich, cites a speech by Himmler: "[T]he Nordic race did not evolve, but came directly down from heaven to settle on the Atlantic continent" (Kurlander 2017: 187). Himmler attempted to square the circle – making the Aryans both ancient *and* modern. They did not need to evolve because they were born perfect. Thus, the Nazi extermination of Jews was not considered a selfish act but was seen as a part of some altruistic grand design – the saving of nature from the "vermin."

The amphibious beings that appear throughout *Hellboy* comics are likewise presented as both past and the future; what was and will be. They have ties to the primordial history of the earth. Their closeness to the sea is, therefore, another signifier of their purity. However, Abe was not born this way, he was transformed. Abe knows his history and that he is not "pure." He rejects the very notion that he could be marked as special for what he is rather than what he does. This is a recurring theme in *Hellboy* stories, as recognised by scholar Scott Bukatman: "[T]he overarching narrative of *Hellboy* is, after all, about the character's defiance of his programming. Hellboy emerges as a character with will and agency, despite the constancy of the forces that try to seduce him into fulfilling his 'destiny'" (Bukatman 2016: 79). The antagonists in these stories often view the protagonists through the lenses of "fate" thus ignoring the individual. Because Abe Sapien is amphibian he must, according to the logic of the Oannes Society, go with their plan.

While the Society views his metamorphosis as a re-connection with a mythical past, a fulfilment of destiny, Abe begins to view it as a chance at evolution. He specifically rejects his past as Langdon Everett Caul, becoming something, and someone, new in the process. This is the greatest of crimes in the eyes of Nazi ideology (and of the Oannes Society): to reject one's connection with the past, to throw one's lot with "lesser" individuals who are not amongst the chosen. Abe Sapien has friends who are, emphatically, not like him. Furthermore, he cares for the fate of all humanity, not just of his "own"-group. Abe refuses to let "fate" make choices for him, to be only this symbolic figure. Abe chooses to be an individual, a human (and humane) being, rejecting the tenets of ecofacism.

Making Movies

Abe Sapien is also a major character in the two *Hellboy* movies (*Hellboy* (2004), *Hellboy II: The Golden Army* (2008)) by director Guillermo Del Toro, played by actor Doug Jones.[2] Though his backstory is not explored,

2   In the first film David Hyde Pierce gave the voice, but all the physical acting was done by Doug Jones.

the films still play up the theme of a world in transition and what will come after. Abe's role is especially pronounced in *Hellboy II: The Golden Army*, in which he falls in love with Elven princess Nuala (Anna Walton). Her brother, prince Nuada (Luke Goss), serves as the antagonist. He wishes to take over the titular army and use it to kill all of humanity – leaving the Earth to him and his kind. Nuada's motivations and dialogue are pure ecofacism. He wishes to kill humanity so that he could save the planet for those that deserve it more.

Nuada tells Hellboy (Ron Perlman) that he belongs with their special group, that he must renounce his humanity to survive. Later in the film he unleashes a monster within a city and forces the protagonists to choose between saving the city or killing the creature: "Look at it. The last of its kind. Like you and I. If you destroy it, the world will never see its kind again [...]. You have more in common with us than with them" (del Toro, 2008). It's important that Nuada creates an "us" and "them" mentality, which is vital for an ecofascistic thought. Garrett Hardin and Pentti Linkola, two prominent figures in the field of ecopolitics, use the analogy "lifeboat ethics" to justify whatever actions they needed to take to save lives: "When the lifeboat is full, those who hate life will try to pull more people onto it, thus [causing the boat to sink] drowning everyone. Those who love and respect life will instead grab an axe and sever the hands clinging to the gunwales" (Linkota 2011: 130–1). Note the framing: the violent action has been forced upon them, allowing them to kill without blame.[3] The idea that one must prepare for possible drowning by having more than one lifeboat is never brought up in this thought experiment.

Hardin did not make much effort to hide the racist nature of his thought experiment. Comparing the United States to a swamped lifeboat and discussing which type of immigrants the good ship United States should take: "My position is that this idea of a multiethnic society is a disaster. That's what we've got in Central Europe, and in Central Africa. A multiethnic society is insanity. I think we should restrict immigration for that reason" (Hardin 1997: n.p.). Like Nuada, Hardin believes in "us" and "them." The death of

---

3   This is tied, once again, to idea of "fate": the ship was fated to drown and people who got on the lifeboat first are inherently more worthy to survive – otherwise they wouldn't be fated to be on the boat.

"them" is presented as a natural by-product of logically necessitated action, but it is *actually* the desired result.

Nuada's plan is similar to that of The Oannes Society. They both conceive the murder of millions as a mere acceleration of a pre-determined process: "Man is doomed, Caul. The air is dark, the seas are sick, and war is coming" (Mignola and Arcudi 2007). The ship is already sinking, and what they do is merely to sever the outstretched arms pleading for salvation. That they helped the boat sink, that they chose who deserves to survive, that they expect to lead this new world is a mere trifle. Those who speak in terms of "lifeboat ethics" only rarely expect to find themselves stranded in the open sea – it is they who will do the stranding.

Of course, as an amphibious being Abe would do quite well in the question of lifeboat ethics – she can survive either on or off the boat. In the film Abe implicitly tells Hellboy that he loves the princess because he considers her to be similar: "She ... she's like me! A creature from another world ..." (del Toro 2008). Like him she could chose to survive while letting others die and like him, she refuses. The film allows its protagonist to confront, each one in turn, the tenants of ecofascism only to reject them.

Brave New Worlds

The figures who consider themselves heralds of the new world (be it Nuada or the Oannes Society), lack the emphatic understating of the cost. They see themselves as one thing only and can thus ignore everyone who is not the same as they are. Their good is the universal good, because they are the universe. As an amphibious being, as a man transformed into something else, Abe is the scion of two worlds – which means he can view them both with equal amounts of empathy. As the *Abe Sapien* story progresses, Abe himself mutates farther away from humanity the closer he clings to it, in spirit if not in form (see Figure 52). As one antagonist asks: "If he's the new man, if he's the future, why can't he accept that this was meant to be?" (Mignola at al. 2013).

Figure 52. Image from *B.P.R.D.: The Dead* #4 (Mike Mignola, 2005). Copyright Dark Horse Comics reproduced under Fair Use legislation.

In the story *B.P.R.D.: The Dead* Abe encounters the ghost of Caul's wife, who offers him to live in an eternal simulation of his old human life: "Separate from the world. And beyond the reach of time" (Mignola and Arcudi 2007). He refuses her. Abe accepts his transformation, accepts that Caul is dead.[4] However, that does not mean Caul has disappeared complexly, he is still a part of Abe, if only as a memory. Abe cannot accept what was "meant to be" because to do so would be to let go of the most human thing of all – freedom of choice. To choose to resist fate becomes the very thing that defines Abe Sapien, as it has defined Hellboy before. Just as he has chosen to be Abe, and

4 · The title of the story refers to several plotlines –all involve dealing with the aftermath of death.

not Langdon Everett Caul, not to be human or monster – but something in-between.

By making that choice Abe also makes a second, greater, choice. To take responsibility not for himself but for others as well, to pull those downing unto the lifeboat, to make sure the world never drowns in the first place. Not to accept the world, or fate, as is but to try and make things better.

Kevin J. Wetmore, Jr.

Moana (Ron Clements and John Musker, 2016)

"In the beginning, there was only ocean" (Clements and Musker, 2016). With these words, Grandma Tala tells the story of the people of the South Pacific and begins Disney's film *Moana* (2016). From the ocean emerge the gods, Te Fiti, the mother island, monsters and everything else found under the sky. The world of Polynesian culture begins with ocean. Grandma Tala tells stories and prepares Moana to rebel against her parents and become a wayfinder, returning the people of Motonui to their status as a seafaring people after staying too long on a single island. As with most Disney films, historical and cultural accuracy are sacrificed for storytelling and to make the narrative family friendly. For example, in the film the demigod Maui claims to have killed an eel and buried "his guts," which is where coconuts come from. In actual Polynesian narratives, Maui sleeps with the wife of the Eel King and when discovered, subsequently castrates him and buries his testicles in the soil, thus creating coconuts, an aetiology a bit too mature for the Disney crowd. On the other hand, the film does capture a moment known to anthropologists as "The Long Pause," a period of two thousand years during which Polynesians halted all ocean exploration. One theory why wayfinding began again is accurately suggested by the film – an ecological disaster made it necessary for people to abandon islands due to a now-limited food supply (Grimaud 2021).

Moana thus transforms Polynesian understanding of ocean into a more contemporary, American, family-friendly version as well. The ocean in *Moana* is a character, a sentient being that has agency and an agenda. Oddly, despite the importance of the ocean to the Polynesian and Oceanian world (right there in the name of the latter, really), the ocean itself is not a sentient being in Polynesian traditional culture. *Moana* transforms the ocean into a character as part of its larger narrative and decontextualisation of the story the film tells.

The cultures of Oceania, while containing many commonalities, are not a monolith. Different islands have variants in the traditional narratives and beliefs. There is overlap between the Hawaiians, the Samoans, the Tongans, the Māori and the other peoples of the South Pacific from Fiji, Rapa Nui, the Cook Islands, the Marshall Islands and many others. They have a fairly common language with local variants, and many common cultural elements (Callicott 1994: 189). *Moana* draws from all of them. The ocean, as Richard Craig (2004: 190) observes, is one of the most significant elements in Polynesian life – provider of food, entertainment, transportation and transit, occupation and material culture – common to all Polynesian culture. The Polynesian people see the ocean as the "great highways on which they easily sailed from island to island" (Craig 2004: 190). What they do not see is the ocean as a being unto itself.

In *Voices of the Wind: Polynesian Myths and Chants*, Katharine Luomala records a chant which proclaims: "The sea belongs to The Lord of the Ocean" (Luomala 1986: 28). In other words, there is a being that is the Lord of the Ocean, but the ocean itself is not a sentient being or a god. Samoans and Tongans have no ocean creation stories; the ocean has always been and will always be (Craig 2004: 191). In other words, the ocean is not personified – it is the universe itself. On the other hand, for the Māori, the ocean is female. Her name is Hine-moana – Moana, specifically, is the Polynesian name for the Pacific Ocean (Craig 2004: 190–1). In Māori traditional belief, Hine-moana is at war with Mother Earth, who is protected by Hine-one (Sand Woman), Hine-tuakirkiri (Ground Woman) and Rakahore (Rocks) (Craig 2004: 191).

All of which is to say that the Disney film is radically different than its mythic and historic inspirations. Grandma Tala's (Rachel House) opening story narrates not just the origin of the world and all that is in it, but also establishes relationships between those elements and provides aetiologies for why the world is the way it is, while providing a path forward for the ecological challenge facing the village of Motonui. Te Fiti created the land, but Maui (Dwayne Johnson), a trickster demigod, stole the heart of Te Fiti, a glowing green gem. As a result, Te Fiti, the nurturing mother goddess represented in greens and blues and with a kindly face is transformed into Te Ka, an angry lava monster. The two sides of the earth are thus captured in the same divinity – kindly, green nurturing mother and dangerous, lava-spewing entity that burns and destroys all it touches. Te Fiti sleeps on the ocean, Te Ka is damaged by it.

In the absence of Te Fiti's heart, the world begins to grow corrupt and the environment begins to fail. For the village of Motonui especially, the coconuts are diseased, the waters around the island are all fished out and the resources are dwindling. Moana (Auli'i Cravalho), the daughter of a chief, will be the leader of the village herself someday. She shows herself a competent leader of the community, but she herself remains drawn to the ocean since birth, an obsession her father, Chief Tui (Temuera Morrison), disapproves of because his best friend died, and he almost drowned when they tried to sail "beyond the reef" (Clements and Musker, 2016). Since the ocean is a dangerous place, the people stopped using it as the highway and stopped voyaging beyond the island.

As a child, Moana rescues a recently hatched baby sea turtle attempting to cross the beach as gulls seek to eat it. Moana uses leaves to protect the turtle until it reaches the waves. Seemingly in response to this, a sparkling wave of energy moves through the water, indicating something has happened and the water has changed. The Ocean[1] then moves and shapes the water to allow the toddler to walk out into the bay without getting wet or drowning. The Ocean shapes a channel of air to allow her to walk along the bottom and witness the ocean from within. She finds a conch shell, and the ocean rises up in a wave that matches her gestures and movements. It plays with her as a mirror image. Moana, "ocean" is mimicked by the moana, the Ocean. The two are connected. Hearing her parent's approach, the ocean picks her up and gently carries her to shore, depositing her safely on the sand, establishing the link between character and water. It is our first indication that the ocean in sentient.

While her father declares the Ocean as "dangerous," Grandma Tala informs Moana that the ocean is "mischievous" and that it "misbehaves," both constructions (dangerous/playful) suggesting not a natural phenomenon but an entity with behaviour, attitude and agency. Whereas her father suggests that the village and the island are all she needs, Grandma Tala reminds Moana, "We have forgotten who we are" (Clements and Musker 2016). The people

---

[1]   From this point forward I will use the capitalised "Ocean" to describe the character with agency and the term "ocean" for the ordinary ocean upon which the characters voyage.

of Motonui are too land-based, and the Ocean wants them back. One possible reading of the film is that Ocean allows and spreads the environmental destruction caused by the loss of Te Fiti's heart not because they are enemies in Māori culture, but because the people have become too dependent on land which cannot sustain them. The Ocean has selected Moana as a saviour figure to return the people to a more sustainable life of wayfinding.

Indeed, at the end of the film, Maui informs Moana:

> I figured it out. You know, the Ocean used to love when I pulled up land, 'cause your ancestors would sail her seas and find 'em. Find all those new lands and new villages. It was the water that connected them all. And if I was the ocean, I'd be looking for someone like you to start that again. (Clements and Musker 2016)

Maui states the point and purpose of Ocean – it exists not just in and of itself, but to serve the people. If the people no longer sail the ocean, it causes not just a problem for the ocean but for all creation, the land as well. Land and ocean are not enemies but must work together to provide a balanced environment in which the people can thrive. Ocean is anthropomorphised to serve an anthropocentric universe in this film. Makes perfect Disney sense, but it does represent a distortion of Polynesian belief.

As she lay dying, Grandma Tala tells Moana she was there the day "the Ocean chose you." "Why would it choose me?" Moana asks (Clements and Musker, 2016). The film spends the rest of its duration demonstrating how the Ocean is sentient, and Moana is its chosen saviour for the people of Motonui. Yet, undercutting this construction of Ocean as powerful sentient being with an agenda is the fact that the film also transforms the Ocean into a typical comic secondary helper character found in Disney films – think the genie in *Aladdin*, Mushu in *Mulan*, or Sebastian in *The Little Mermaid*. Rather than the all-encompassing universe of Polynesia culture, *Moana* reduces Ocean to a comic sidekick, which if it had lines would have been voiced by a famous, popular comedian. When Moana falls asleep at the till, Ocean comically taps her on the shoulder. Frustrated at her inability to locate Maui the demigod, she cries out: "Ocean, can I get a little help?" (Clements and Musker, 2016), and immediately a terrifying storm rises, depositing her on Maui's island.

Once Maui steals Moana's boat, Ocean manifests as powerful force and comic presence. Ocean propels Moana through the water and onto the boat.

When Maui throws her off the boat, the Ocean places her right back on deck, dripping wet. This act is played for laughs by repeating several times – Maui throws Moana off the boat; Ocean puts Moana back on the boat. Maui jumps off the boat; Ocean puts Maui back on the boat. Heihei (Alan Tudyk) the chicken falls off the boat; Ocean puts Heihei back on the boat. Heihei falls off the boat for the third time in a row; Ocean puts Heihei in a storage compartment in the hold and then closes the hatch so he cannot fall again. Ocean is an unstoppable force with a sense of humour. Grandma Tala, we see, was right – Ocean is mischievous and funny.

Moana takes her role as Ocean-appointed saviour seriously. When Maui asks why her people sent her, she responds: "My people didn't send me, the Ocean did." Maui's response: "The Ocean is straight up kooky dooks" (Clements and Musker 2016). While Maui's response is indeed comic and dismissive, like the humans he knows the Ocean has a personality, agency and agenda.

When she is discouraged over her own inability to wayfind, and Maui tells her their inability to restore the heart of Te Fiti is due to the fact that "[w]e're here because the Ocean told you you were special and you believed it," (Clements and Musker 2016). Moana rejects her role as saviour. Maui's statement is remarkable in that it asserts that Ocean is indeed sentient, but it lied to Moana. The problem is not that Moana has delusions of grandeur, but that she believed the Ocean. That aspect, however, has been her quest all along – believing the Ocean and working to restore the people to wayfinder status.

At her low point, she rejects the assertion and throws the heart of Te Fiti back into the ocean, telling Ocean to choose someone else. Never is the Ocean's ability to choose or have agency in the world questioned, only its specific choices. Interestingly, Ocean accepts Moana's choice. We know it can return the gem to her at any point. We have seen its power to insist on getting its way. Instead, Ocean allows the gem to sink to the bottom of the ocean. After Grandma Tala's spirit manifests in the form of a large manta ray, Moana realises she must follow through on her commitment to restore not just the heart of Te Fiti, but the Ocean itself as the heart of the Polynesian world. She dives in and recovers the heart, making her way to the island of Te Ka.

Recognising as they struggle that Te Ka is simply Te Fiti without her heart, Moana instructs Ocean to "[l]et her come to me" (Clements and Musker

2016). Just as at the beginning of the film we see Ocean clear a path for baby Moana to walk out onto newly dry land in the middle of the ocean, Te Ka is able to cross to Moana, who restores the heart. While the restoration of the heart has a curative effect on the island of Motonui, the restoration of the land is no longer as necessary, as Ocean has now also been restored as the centre of the people's lives.

The film ends with the people of Motonui once again ocean-travellers, led by the wayfinder Moana. In short, the Disney film does not accurately depict Polynesian belief, but it does present a comic, anthropomorphic, sentient Ocean that works with demigods and human girls to restore balance to the world, end the "Long Pause" and return the people to the ocean and the ocean to its right place in the world. Ocean uses Moana as a saviour figure to rebuild the world, and in doing so serves its own needs and restore its own primacy, even if it is kooky dooks.

Ruth Barratt-Peacock

Ocean Poems (David Malouf, 1976–1991)

Australian poetry is as diverse as the life-worlds and interests of its authors. However, its reception in academia tends to revolve around place and place-making in (post)colonial contexts. Whether in the city, suburbs, bush, desert, or "nature" more generally, Australian literature has traditionally been viewed as struggling with writing a lasting connection to land, which is due to the (ongoing) conditions of colonisation (see Stilz 2006; Wright 1965; Rigby 2009). The conundrum of Australian literature can be summarised as a land/language disjunct (Wright 1975: 59). Oddly, for an island continent, the ocean remains underrepresented.[1] Yet, the ocean offers a valuable alternative way of thinking about the human/Other relationship precisely because of its instability and fluidity.

In the poems of multi-award-winning Australian author David Malouf, whose work I focus on in this chapter, transcendence is figured as a post-human connection to the animal other. The ocean is key to this vision of transcendence because it offers no ideal and final unity of human and Other. Transcendence in the ocean is not a state to be reached, but an endless process of reaching for the Other. By failing to reach a state of post-human consciousness outside language and by reflecting on this failure in language through art, the human subject gains an awareness of the limitations of human knowledge. This is particularly important now, in the context of the Anthropocene, in which scale effects make it almost impossible to truly grasp the culminative effects of everyday actions. As Clarke writes, "we have a map whose scale includes

---

1 Related sites, such as the island or the beach as a site of colonial meeting, violence and survival, are not considered part of the ocean, particularly part of the deep for the purposes of this chapter (see Crane 2019; Fiske et al. 1987; Fletcher and Crane 2017).

Becoming Ocean, Becoming Self

the whole Earth but when it comes to relating the thread to daily questions of politics, ethics, or specific interpretations of history, culture, literature or other areas, the map is often almost mockingly useless" (Clarke 2015: 71). This poses a unique challenge to cultural products and how we interpret them in academia. To put it another way, "the Anthropocene is a crisis of ontology which brings with it an epistemological crisis the humanities are scrambling to catch up to" (Barratt-Peacock 2019: 180).

The following focuses on three of Malouf's poems: "In the Seas' Giving" (1976), "The Crab Feast" (1981),[2] and "Pentecostal" (1991). These poems show a desire for a connection with the Other and the deep-seated fear that this is not possible. Malouf's work goes back four decades, well before the appellation of the Anthropocene as a new geological epoch. Nevertheless, there are notable similarities to recent ecocritical thought, in particular Rigby's call to lay aside our reading and "seek out the kinds of embodied experience of the living Earth that might strengthen us in our desire to defend our Earth others, [...] a desire whose fulfilment nonetheless necessarily returns us once more to the world of words" (Rigby 2009: 177). In the poems explored here, we see an exploration of the embodied experience of the Other and the problematisation of the world of words itself.

The fundamental problem with any re-thinking of human subjectivity is one of language. To imagine a post-human self is to think, and thinking entails the use of language. However, language is that which first gives the thinking subject a sense of self (the proverbial *cogito ergo sum* of Enlightenment dualism). The subject is aware of itself only as a being separate from others, including the animal Other. Reading Malouf in the Anthropocene, we can see an implicit and damming commentary on an idealisation of "communing" with the non-human, to transcend the human in much green-thought (see Taylor 2010). Meeting the Other will always be a limited and violet affair.

---

2   From the collection *First Things Last* (1981).

Diving in the Deep: Imagining Transformation at the Foundations of the Self

The ocean is familiar and strange at the same time. In "In the Seas' Giving" it is described as a "world so close that was not/ours. Though we could swim, / my sister and I, / before we could walk" (lines 1–4). The uncanny or *unheimlich* ocean is essential because it allows the subject to assert their place in the ocean as something known and point towards the unknowable Other in relation to that. It is the desire to find the Other by diving down to the foundations of the self. In "In the Seas' Giving" this occurs in the form of diving into the ocean. The deep is figured as the primal womb of humanity by linking it to the waters in the womb surrounding the foetus before it is born and learns to walk. The subject seeks the foundations of the self in the unconscious (situated in the ocean through this imagery). The impossibility of this is made productive in conceptualising a post-human future, or rather, in pointing out the limits of language in imagining such a thing. The poet reflects on the poem's own inability to create an experience of non-human consciousness in the ocean by noting that the children *fail* to "make friends / with its creatures" even "in dream" (lines 11–13). The speaker desires to find kinship with the Other of the ocean in their own unconsciousness, as shown by the following lines from "The Crab Feast": "The back of my head / was open to dream / dark your [the crab's] body moves in" (lines 78–80). However, like reflections on the water and dreams upon waking, it cannot be successfully integrated into the daily life of the human subject.

The failure of the imagination to unify the swimmers with the animal Others in "In the Sea's Giving" is highlighted in the seventh stanza (see Barratt-Peacock 2020: 93–4), where "These others / melted in our heads" once "Hauled out into the sun" and the humdrum world of the poet (lines 25–7). This humdrum world contrasts with an imagined, wilder past, in which the human was hunted like the sea creatures that the speaker now hunts. Importantly, the poem does not fix these relationships through statements, but reflects the fluidity of the ocean in its form, leaving the reader with a question:

What / had we traded to be safe? / Past midnight, sealed off / in the bronze bell of the moon, / we plunged over the sea ledge seeking / answers, climbed back / at dawn wearing our magic / skin – only later / found what we had lost. (lines 32–41)

As my previous reading of this piece explores, this question points to "the loss but also the result of this loss" (Barratt-Peacock 2020: 94). What has been lost is pre-linguistic unity with the world and its animals. When the subjects of the poem do find it "only later," it is not in a fleeting experience of the ocean, but in the dissolution of the self *as* the ocean. The "ghosts, fish-like, transparent" which float out of the subjects are the animals of the ocean merged with the human (line 44). This more-than-human experience is lost again when returning to the shore. Such renewed loss occurs in the moment that the subjects reflect on the experience and there is the bind: this knowledge requires reflection, and "can thus only be attained in the negative, as an understanding of what is absent" (Barratt-Peacok 2020: 95). The absence is the absence of human linguistically-determined consciousness and it is not surprising that this is tied to death. The question is, whose?

Communion and Transformation through the Flesh

Kinship between the human and the aquatic other in the ocean is asserted many times in Malouf's poetry, with examples seen in the lines from "The Crab Feast": "[W]e were horizons / of each other's consciousness" (part 6, lines 1–2) or "You were / myself in another species" (part 10, lines 20–1). However, this kinship can only become known through a sacrament that penetrates physically, sensuously and imaginatively into the Other, through the meal, figured as what Tulip terms "a sacramental moment" (Tulip 1981: 396). The speaker attempts to overcome the human–animal binary through the meal figured as communion both in "Pentecostal" and "The Crab Feast." "The Crab Feast" is a ten-part epic that was written long before the pollution of the oceans following the Fukushima or the rapidly progressing death of the Great Barrier Reef (Barratt-Peacock 2019: 183), yet the dynamic between the human subject and the oceanic in this poem is already a play of violence and the sacred that can be seen in lines such as:

> This hunt / is ritual, all parties to it lost. Even the breaths / we draw between cries / are fixed terms in what is celebrated / the spaces in the net. (part 3, lines 20–4)

There is little room to break out of the hunter–hunted relationship in this piece. The whole poem plays with macro and micro viewpoints to try to establish unity between the speaker and the crab. However, the terms are "fixed" regardless of the ritual celebrated. There is space (even a planetary perspective), but it is only the spaces that define the net which the crab will eventually be caught in. The closeness of the net connects to the unity of subject and Other achieved through eating (and killing). At the same time, the spaces attempt to write a common point of origin for both subject and crab in long history, like "In the Seas' Giving," characterised by a pre-conscious connection in the deep. "The Crab Feast" attempts to merge the self with the other through both a macro and a micro view. The micro lens depicts a sexualised, deathly corporeal desire for merger with the oceanic other: "There is no getting closer / than this. My tongue slips into / the furthest, sweetest corner / of you" (part 1, lines 1–5). Simultaneously, the broader context of this relationship is shown from a planetary perspective that eventually sees their meeting only in death: "[...] to that perfect / primary death colour, out / into silence and a landscape / of endings, with the brute sky pumping red" (part 4, lines 18–21) and "I watch at a distance of centuries, in the morning / light of another planet / or the earliest gloom [...]" (part 5, lines 1–4). Meeting the Other requires a death of the self. A post-human self cannot be imagined in human-linguistic terms. It exists only in silence beyond the sign or perhaps in the physical merger between human and ocean Other purely in the flesh.[3]

---

3   For a more detailed analysis of this poem see Barratt-Peacock 2019: 182–92; Barratt-Peacock 2020: 95–101).

Communion

"The Crab Feast" attempts to re-write eating the crab as a sacred ritual between two equal parties. The conceit of this becomes humorously explicit in the poem "Pentecostal," which recalls Christian symbolism with a mixture of earnestness and satire (Barratt-Peacock 2020: 101). The title is already a play on the Pentecost:

> Lit scales in the grass, in a dented pail soft guts / that spill their lost light thin as the Pleiades. / Mullet, whiting, bream – the terror of Fridays / their names, their boneless ghosts, lodged in my throat. / [...] Gelatinous thick / tongues in a shell too far from the sea, they thunder / Te Deums to table-salt, writhe in the anguished / hiss of deliquescence. I make myself small, / then smaller; approach the vanishing point; surrender / breath then blood to hear from the wet sizzle / of flesh made dung, scoured shell, sucked bone, the clear / death-aria in my throat, the voice of the word-less creature in us that rages against hook, / against withering salt, against frying pan and fire. (lines 1–4, 9–18)

Describing the sea creatures being eaten as "the terror of Fridays" recalls the childhood horrors of having to eat fish in line with the Christian tradition of eating fish on Friday. The poem evokes the idea of sacred sacrifice and communion through the fish because fish is a fasting food eaten in honour of Jesus' sacrificial crucifixion in the Christian tradition. At the same time, the implication of a childhood recollection and descriptions such as "flesh made dung" or "gelatinous thick tongues" brings in an element of satire and disgust undermines the speaker's attempts to deny their selfhood and dissolve into unity with the Other. The sea creatures in the bucket are visceral and dying a mundane death in the kitchen. The imagery offers an almost satirical take on the tongues of flame in Acts 2–4 (that allowed the disciples to speak all languages) as "Gelatinous thick / tongues," as well as Jesus as the Word made flesh in "the wet sizzle / of flesh made dung" (lines 14–15; Barratt-Peacock 2020: 102). Language is, as always, key, and the poem is an attempt to transcend language in many ways. However, death and language are shown to go hand in hand. The sea creatures are given language, but only in a mockery of the sacred as they chant "Te Deums" (a song of praise) to the "table-salt" in which they are about to be smothered. The salt of the ocean, become a salty death. The speaker surrenders their breath to the Other, but only to find the need to also surrender their body, in the lines "surrender / breath then blood." These lines imply that even though physical surrender of the human subject to the animal makes it possible for the human to experience a kind of unity with the oceanic Other, this only happens within the framework of death and violence, however much the subject wishes to imagine it otherwise.

Conclusion: The Deep and the Eco-Self

More than simply a matter of naming a new geological epoch, it can be argued that naming our age the Anthropocene "demonstrates that morals, culture, and geology after all have something to do with each other" (Parikka 2015: 7). The material and the cultural are intertwined and, in Malouf's sea poems, it is a deadly embrace. Language cannot facilitate a post-human future; it cannot even imagine it without evoking death in a way that "would seem to work against any notion of a surrender of control by a dominant human species" (Smith 2009: 167). Ecocriticism calls for a radical re-thinking of the human. In the now classic text *The Companion Species Manifesto: Dogs, People, and Significant Otherness,* Donna Haraway points out that beings "don't pre-exist their relations" (Haraway 2012: 6). The human and the animal are co-constitutive, both biologically and socially. Falconer expresses something similar, saying that "the kinds of animals we've witnessed, and lived around make us different people" (Falconer 2017 qdt. in Barratt-Peacock 2019: 177). The final line in section 7 of "The Crab Feast" makes a lot more sense in light of these ideas: "Because you are so open / because you are." "Because you are" implies "because I am." The nature of the crab is determinative of the nature of the subject and neither pre-exists the other. Imagining a post-human self that has overcome the self–Other divide is not only impossible but misses the mark. The subject enters the ocean to meet, not the other, but the self in the ocean. The ocean is the perfect reflection of the foundation of the eco-self: there is no foundation, only movement.

Octavia Cade

Évolution (Lucile Hadžihalilović, 2015)

The French ecological horror film *Évolution* (2015), directed by Lucile Hadžihalilović, locates cross-species reproduction in the shifting waters of the intertidal zone. This coastal imagery, shot from both above and below the waters, links the human and non-human species of the film in a way that perfectly exploits the dual nature of the setting. That the intertidal zone is a place of liminality underlines the transgressive nature of the text, breaking down biologies and offering a glimpse of a world where neither gender nor species is a barrier to motherhood.

The intertidal zone is notable, from an ecological perspective, in that the species that survive there are profoundly adaptive. They have to be: a species that can survive periodic immersion in seawater, and which is subject to an ever-changing tidal position, must be able to endure extremes of temperature, desiccation, salinity and insolation, amongst other environmental factors (Dring 2003: 35). That the events of *Évolution* are so often centred on the intertidal zone is an indication of the inherent ability of the characters to navigate both marine and terrestrial ecologies.

This navigation initially appears ambiguous. Nicolas, like all the other young boys of the island community of *Évolution* – and the children are only boys, as the adults are only women, with each representing only a single generation within their age groups – is unwell. His time is divided between the hospital and the beach, where all the children are regularly taken by their mothers in order to bathe. This is reminiscent of the seaside cures of historical times, when patients suffering from a number of maladies were advised to bathe in ocean waters to improve their health. The first indication that the collective sickness of the children is not what it seems is in their general appearance, which is generally fit and healthy. By contrast, the mothers are collectively pale and washed-out, even though they spend as much time sea-bathing and in the intertidal zone as the children do. This may, of course, be their natural state. If not, however, then their washed-out appearance is only to be expected, as different species exhibit different tolerances to environmental stressors, and the mothers are all adoptive as well as adaptive: it is only the children who are fully human.

The mothers, in *Évolution*, are sea creatures. They do not easily align with any particular legend or mythology, such as mermaids or selkies. Rather, the most obvious visual characteristic of their difference lies in the suckers scattered over their backs. These suckers are initially reminiscent of cephalopods, such as squid or octopus, but the continual imagery, throughout the film, of starfish, indicates an alternate possibility. Starfish also have suction cups on the undersides of their arms, although the tubular shape of the cups does not truly reflect the physiology of *Évolution*'s mothers. There is no compelling reason, however, to map those mothers to a single marine species. It is enough that they exhibit physiological characteristics that are recognisably marine in origin.

If different species exhibit different tolerances for environmental stress in the intertidal zone, however – and it should be noted that there are species of both starfish and octopus that live in the intertidal zone – then some species are more suited to different positions within that zone. If the isolated island community of *Évolution* lives primarily by the sea, all the buildings, including houses and hospital, are located in solidly terrestrial environments. They may border the beaches of the island, but their primary purpose lies in providing stable homes and medical care for the adopted (human) children. It is no wonder that those children are better suited for the terrestrial environment above the intertidal zone, even if the intertidal zone is where they spend most of their waking, non-hospitalised hours. Their mothers, spending more time than might be wished in the upper reaches of the intertidal, and in the coastal areas above it, are potentially less adapted to the environmental conditions there.

A secondary factor can be observed in the vertical distribution of intertidal organisms that may shed light on the complex species relationships of *Évolution*. A generalised characteristic of species distribution in the intertidal zone is that their upper distribution is primarily influenced by abiotic factors such as light and temperature, while their lower limit is more heavily influenced by competition and predation (Dring 2003: 132). To give an example, an intertidal starfish may dry out too much if it travels too far up the intertidal zone, while the further it moves into the water, the more likely it is to suffer predation from other marine organisms. *Évolution*'s mothers may be limited in their movement away from the ocean and the intertidal zone, but the closer they get to the sea, the stronger and more predatory they become. Certainly, they are competing with humans for resources, particularly reproductive resources. Nicolas has dim recollections of life before he came to the island, drawing a picture of a Ferris wheel; this implies that he, like the other boys, was at some point stolen and relocated to the island.

These children are raised by their adoptive mothers, but the uncanny behaviour of those mothers – mimicking human relationships, if not entirely human form (if indeed their form is a mimicry, a form of camouflage, instead of their own unique physiology) – raises suspicions in Nicolas at least. And indeed, when apart from their children, distinctly non-human behaviour begins to emerge. In one scene, Nicolas sneaks out of his bedroom at night to follow the mothers as they make their way down to the intertidal zone. He observes a naked, almost orgiastic writhing in the wet sand, and it is surely no stylistic coincidence that this group consummation begins with the appearance of a starfish, as the mothers position themselves into a replica of that form. (Images of starfish, and images reminiscent of starfish, are continually presented to the audience, whether in the form of a child's drawing, or the position of lights on a medical lamp. This is, surely, not deliberately imitative on the parts of the characters, but it is nonetheless reflective of the slow colonisation of the island by the sea creatures.)

This night-time ritual, practised by the mothers, is arguably a characteristic of their ecology and behaviour kept both separate from, and secret from, their adoptive children. For the boys to tolerate the constant medical experiments performed on them, they must accept that they are genuinely sick, and that their mothers have their best interests at heart. Those mothers do not. Their primary concern is reproduction, and in the lower, more competitive regions of the intertidal zone, that reproduction goes hand in hand with predation.

Interspecies reproduction can occur without technological interference, as proved by the existence of lion/tiger hybrids, for example, but the mothers of *Évolution* appear incapable of natural reproduction. The film does not address the causes of their infertility; it is largely silent on many aspects of the mothers' biology, with an extremely minimal script. It is possible for viewers to speculate, but not to confirm. Perhaps this non-human species once included males, but those males became extinct. Perhaps the warming waters of a world altered by anthropogenic climate change has led to an opposite sex distribution to that observed in some fish, where warming waters can result in a disproportionate number of males within a species (Honeycutt et al. 2019). Perhaps the sea creatures are shifting from asexual to sexual reproduction. Perhaps the mothers are capable of interbreeding with adult human males but simply choose not to do so. For whatever reason, the burden of reproduction, in this population, has fallen on young human boys.

Those boys are artificially impregnated. Injections into their abdomens are followed by scans such as those performed on pregnant women, and the most warmth that Nicolas' mother ever shows him is the smile when she hears the heartbeat of the creature living inside him. At a late stage in the pregnancy, the boys are restrained in tanks full of water, with only their heads above the surface, and Nicolas wakes at one point to find two small creatures attached to him and feeding off him.

Clearly there is no question of informed consent. The children, who are around 10 years old, are incapable of giving that consent, and their objections would not be listened to anyway. Furthermore, it is also clear that many of the boys do not survive their forced pregnancies, and many of the embryos do not either. There is even a storage cupboard full of foetuses in jars, with a disturbing mix of human and marine characteristics that have clearly resulted from the equivalent of an early miscarriage.

If the environment that the boys are raised in is indicative of the upper distribution of species in the intertidal zone, then the fundamental relationships between the two species in *Évolution* are more suited to the restriction surrounding the lower vertical distribution of said species. The sea creatures of *Évolution* appear to be a dying people. They can no longer reproduce on their own, and their remnant population can only survive on an isolated island where there is little competition from the dominant (human) species. Furthermore, they resort to reproductive strategies that are entirely dependent upon the predation and exploitation of their rival species' young. The mothers of *Évolution* may not be literally eating their adoptive offspring, but they are certainly using them as consumable resources by which to preserve their own species.

Nicolas, along with the other boys, is being used as an incubator in order to assist in the reproduction of these sea creatures, and the boys' continual activities on the shoreline underline their status as intertidal beings, being both of the ocean and separate from it, existing in a littoral environment that has more than a passing resemblance to the liminal. This raises questions of identity that are naturally conflicted. Rejection of the maternal bond is emphasised by early indications of antipathy towards other oceanic species, as observed in Nicolas' violent reaction to starfish – he smashes an arm off a starfish with a rock – but this antipathy is undermined by a forceful expectation of nurturing. The damaged starfish is placed safely in a fishbowl and settled in his bedroom, presumably by his mother, the same woman who is complicit in Nicolas' impregnation.

Nicolas may accuse the woman who takes care of him of not being his true mother, but his own maternity is also in question. Is he merely a surrogate for the sea creatures around him, or does part of his own genetic material pass on to his offspring? And given that he survives the pregnancy, does he retain in his own body elements of the sea creature that impregnated him, as human women do? Named after the monstrous hybrids of Greek legend, microchimeras occur when cells from the foetus pass through the placenta and into the mother's body, where they can multiply and accumulate over decades (Callier 2015). Should this same phenomenon occur to Nicolas, he would become chimeric himself, literally part sea creature.

Given the emphasis on maternity in *Évolution*, it should finally be noted that the intertidal zone depicted in the text, alongside the underwater exploration of the subtidal zone that opens the film, is the primary source of non-human life within the narrative. The land above the intertidal zone is visually stunning but almost entirely barren, consisting primarily of bare rocks. A stunted, browning grass, shown only sporadically, is the only featured terrestrial flora, and the only non-human terrestrial animal shown is a single lizard. In comparison, the intertidal and subtidal zones are associated with bright images of an extensive and functioning biota. Different varieties of starfish are shown, as are luxurious algal growths. These act as a reminder that the ocean is the source of life for the island depicted in *Évolution*, as it is in the rest of the world. The genesis of life on Earth is, fundamentally, marine-based, and that life colonised, eventually, the land that bordered the oceans. Arguably, the ocean may be interpreted as a maternal environment, albeit for a maternity that first arose *very* far back; an interview with Hadžihalilović that notes the similarity of *la mer* and *la mère* reinforces this (Romney 2016).

The sea creatures of *Évolution* result from that same maternal environment. In a sense they are very distant cousins, and while there is little in the film to indicate the source of their population decline, we are able, from the images presented, to make certain surmises. The images of the land, when compared to the images of the sea, are bleak and colourless. The land appears sterile; certainly more sterile than the waters of the ocean. Given that *Évolution* is a narrative set in the contemporary world, it is also a narrative set in a world of climate change and significant biodiversity loss. Much of that loss is terrestrial, if only because it is easier to observe and to record ecological decline on land than it is in the sea. If the sea creatures of the intertidal zone are sterile also, or if they become so through the inability of their species to reproduce, even with technological assistance and stolen children, then it is representative of a population and species loss that is spreading through ecosystems, through kingdoms and genera, and through families.

If Nicolas is adopted by a sea creature, raised by a sea creature, is impregnated by and gives birth to sea creatures, and in doing so becomes, on a biological level, part sea creature himself, can he still be described as a human being? Or has he become that monstrous hybrid, an intermediary step in human evolution, and a link between human and non-human populations? That the sea creature mothers are exploiting and endangering their adoptive offspring is certainly monstrous, yet the potential for sympathy exists across species boundaries, as the relationship between Nicolas and the young nurse illustrates. They build a tentative trust on the strength of shared vulnerabilities, as Nicolas shares his drawings and the nurse shares her knowledge, as well as the marine realities of her own sucker-studded body. This indicates that there is a potential for identification that is more than just one way, as interaction with a human child makes the nurse a little more human in her own behaviour.

There is in the seeds of this relationship, then, an argument for recognition, if not entirely one for reconciliation. The increasing impact of human activity on the oceans, and the subsequent inability of many creatures to successfully reproduce in that changed environment, has been well documented. Can these sea creatures be faulted for wanting to survive, no matter the cost of their survival to others? In a world where climate change is having a profound effect on the ecology of the oceans, the ability to effectively navigate questions of agency and reproduction in both littoral and liminal zones is, perhaps, crucial to the survival of more than just our own species.

Justin Wigard

Subnautica (Unknown Worlds Entertainment, 2014)

Envirofuturism can be understood as the practice of humanity reconciling with the environment, constantly concerned about their conjoined
future while dealing with ramifications of the past in the present through
the use of technology, broadly conceived. Michael Gauthier (2018) says
that "the theory or ideology of 'envirofuturism' is founded on the belief
that technological advancement and preservation and proliferation of our
natural environments don't have to be rivals." In this and other myriad
ways, I argue that the video game *Subnautica* acts as a playable instance of
envirofuturism, imparting onto players ideologies associated with preservation and balance between humanity, the environment and their conjoined
future survival. This manifests within *Subnautica*, which is predicated on
exploring increasingly deeper waters on a primarily subaquatic world, and
in doing so, uncovering evermore complex life forms and technologies.
Subnautica provides a playable sandbox to engage with envirofuturism –
to experiment with it, learn about it (indirectly), and apply its ideological
framework within the game.

Unknown Worlds Entertainment's *Subnautica* begins simply and won-
drously: the player – an astronaut in the distant future – crash-lands on the
alien Planet 4546B, a planet almost completely covered by water. From the
first moment of gameplay, the player must stave off thirst and hunger, investigate the crashed ship underwater, build a sustainable habitat and return
home, all while surrounded by alien subaquatic flora and fauna: ever-present
Peepers, one-eyed glowing fish that appear throughout the entire ocean, from
the shallows to the depths; Cuddlefish; Bloodroot; Jellyshrooms (Figure 53),
and more. Early in the game, the player will discover an alien bacterium has

infected everything on the planet, including the player. Simultaneously, the player learns that a planetwide quarantine is the cause of their crash: ancient alien technology destroys any movement on or off the planet, keeping this bacterium locked to Planet 4546B.

The endgame, and the climax of the game's narrative, revolves around meeting, speaking with, and assisting an ancient, gargantuan aquatic being known as the Sea Emperor Leviathan. This Leviathan tells the player that they produce an enzyme that can act as a cure for this planetary bacterium. Three things become apparent: (1) this planet was previously colonised by alien beings who were researching a cure for the bacterium; (2) they experimented on the Sea Emperor Leviathan, hoping to find a cure for the bacterium, and died in the process; (3) the Sea Emperor Leviathan is willing to provide the cure if the player will ensure the survival of its species.

In order to survive, and escape, the player must truly understand the Sea Emperor Leviathan's envirofuturist ideology. Players must come to terms with the flora and fauna of Planet 4546B: not all life can be commodified, destroyed, or otherwise used. Likewise, players learn that the previous aliens'

Figure 53. The player faces off against a looming Crabsnake in the Deepshroom Forest. *Subnautica*, Unknown Worlds Entertainment, 2018. Image credit: *Subnautica* Twitter verified profile.

attempts to colonise the planet backfired spectacularly, an explicit warning to players against such technological practices. Once players agree to the Sea Emperor's mutually beneficial relationship, players develop technology – the cure itself – and distribute it through the Peepers that pervade the planet. *Subnautica* reinforces these envirofuturist ideologies by forcing players to literally embrace envirofuturism as their dominant mode of play, or risk never completing the game's narrative. As a result, players escape and leave the planet behind, resisting colonising practices to inhabit the planet and leaving the space to its aquatic, indigenous lifeforms.

Crash Landing and Crafty Living

Early gameplay in *Subnautica* is brutal on both the player and the environment. Players are immediately assailed with hunger, thirst and injuries as their lifepod is surrounded by an ocean. The larger crashed ship – the Aurora – is the only physical landmark jutting out of water, no land in sight. This means the player must constantly dive to scrounge for resources, using their handheld scanner to learn more about what use each mineral, plant, or animal can provide.

Figure 54 (left). A Peeper, a small fish with a large, globular orange eye. *Subnautica*, Unknown Worlds Entertainment, 2018. Image credit: *Subnautica* Fandom Wiki.

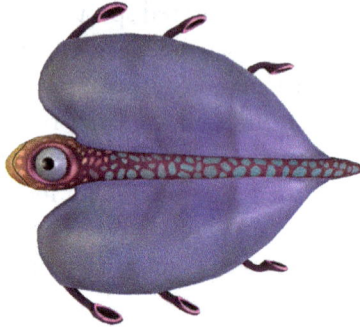

Figure 55 (right). A Bladderfish, a pink fish with large purple lung sacs. *Subnautica*, Unknown Worlds Entertainment, 2018. Image credit: *Subnautica* Fandom Wiki.

In perhaps the clearest instance of this, the player will scan the Peeper (Figure 54), a small fish with a large orange eye that is a wonderful source of food, but they will also scan a Bladderfish (Figure 55), a bright pink fish with large lung-like purple sacs jutting out from its torso. The scanner reveals that this Bladderfish is, indeed, an air bladder fish: it can be captured and brought back to the Fabricator to be converted into fresh, oxygenated water, quenching the player's thirst in an otherwise saltwater ocean. Other instances involve mining gold and copper to be used in conjunction with semi-valuable Table Coral to create a makeshift Computer Chip, among so many other technologies.

What the player discovers is that, while the smaller fish and other aquatic lifeforms seem to respawn at a more or less sustainable rate (i.e. players can never get too many Bladderfish), there is a somewhat limited supply of other resources and lifeforms on Planet 4546B. Further, the Table Coral does not respawn or grow back, meaning it is not renewable through farming or through any in-game respawning. The player then has to find a balance of harvesting the resources they need to build more upgrades, and eventually, to expand their searchable area to new environs. Fortunately, mineralistic deconstruction is among the technological advances apparently made by humans in this game. Essentially, the Habitat Tool can be used to Deconstruct fabrications, returning each mineral or resource used in its construction in a 1:1 ratio. There is no loss of resource once built, ensuring that at least the player's ecological footprint can, more or less, be erased should the player choose.

The game does not incentivise violence, and in fact, it takes concerted effort to actually kill anything larger than the swarming fish due to the lack of actual weapons within *Subnautica*. The player finds blueprints for a knife early on, which the game treats as a tool akin to the Scanner, the Habitat Tool, and others (Figure 56). It is used to kill Bladderfish, Peepers and other fish for sustenance, to be sure, but it is also used to harvest plants, Table Coral and biological samples. To kill something larger than a fish, say, carnivores like the Crabsnake or a Stalker, players must spend several minutes slashing and hacking at the creature. When these larger creatures die their carcasses do not despawn in game parlance – they do not decompose, rot, or otherwise disappear. While these creatures can respawn, their bodies stand as a stark warning against using technology to actively harm the environment. Corpses are grim envirofuturist warnings of using technology – even simple technologies like a knife – against the environment in a harmful manner. Thus, *Subnautica* embeds envirofuturism throughout the gameplay. Players must learn to work with the environment in order to survive, learn its ebbs, flows and inherent technologies.

Envirofuturist Depths

This central game loop eventually brings players to the crux of *Subnautica*'s envirofuturist core. By exploring the ocean, players make three key discoveries. First: they confirm they are not the first extra-terrestrial lifeform to land on Planet 4546B. Second: the previous lifeforms – aliens referred to as Precursors or Architects – tried and failed to colonise the planet, letting loose a deadly plague known as the Kharaa Bacterium running rampant on the planet. Third: something from beyond the player's ken is sending telepathic messages to draw the player ever further into the depths of the ocean.

Relics of these ancient Precursors abound, from sprawling underwater black cables imbued with green light to giant constructions sticking out of the water to research facilities hidden in an underwater volcano. Players further learn that every living thing on Planet 4546B is infected with the bacterium; that the entire planet is under a complete quarantine punishable by surface-to-air laser cannons; and that they are, as luck would have it, infected with the bacterium. It manifests in virulent and phosphorescent green pustules dotting the player's skin, as well as the surfaces of plant and animal life alike. In this way, *Subnautica* throws the player into an envirofuturist state. As players develop a relationship with the environment of Planet 4546B in the present, they must deal with and overcome the sins of the long-dead aliens (i.e. the Kharaa Bacterium) in order to escape the planet. To accomplish this task, players must harvest and repurpose futuristic alien technology for their own survival, building new underwater habitats beyond their cramped escape pod.

Creating new upgrades eventually allow players to work towards building their own self-sustaining habitat. One such technology is the Water Filtration Machine, which automatically desalinates water from the exterior of the habitat, and filters in freshwater for consumption: no more consuming Bladderfish for water! Other sustainable upgrades include plant beds, which allow players

Figure 56. An underwater habitat with the player holding the Habitat Tool. *Subnautica*, Unknown Worlds Entertainment, 2018. Image credit: *Subnautica* Twitter verified profile.

to actually grow their own flora for consumption or use, whether indoors or outdoors. These technologies, of course, take resources from the environment to be created, which takes time to collect and build, but the self-sustainability element allows players to focus on more advanced technologies: underwater submersibles (small and large); more efficient breathing apparatuses; a walking, jumping, grappling-hooking mech suit; and even a rocket capable of breaking out of the planet's atmosphere and into outer space. Each new technological upgrade gives the player a greater capacity for deeper water, until eventually, they are able to explore the planet's depths and meet the being sending these telepathic messages: the Sea Emperor Leviathan.

"This is what the others could not force from me."

Nowhere is the envirofuturist ethos more prevalent than in meeting the Sea Emperor Leviathan (SEL) late in the game (Figure 57). Until the player meets the SEL, they primarily encounter sentient, but low-intelligence creatures, including other Leviathans capable of destroying them. Throughout the game, players gain glimpses into the Precursors' time on Planet 4546B, that they attempted to capture, then exploit, the SEL. This is crystalised when the player meets the SEL deep within an abandoned Precursor prison/aquarium at the bottom of an ocean crater, several thousand metres under the ocean's surface. While the Precursors had captured the SEL in to learn about a panacea-like enzyme she produces, they refused to learn from her, or the environment. Instead, they wielded their technology like a cauterising weapon of destruction for their own doomed gains, rather than understanding the SEL or Planet 4546B. Unsurprisingly, she refused to give the secret of the cure under duress; that is, as an embodiment of Planet 4546B's life and as a steward to her own kin, she refused to impart the biological technology – Enzyme 42 – necessary to save the Precursors from their own demise. They refused to heed the envirofuturist call: their past sins of rampant colonialism caught up to their present, all but eliminating their future.

Figure 57. The Sea Emperor Leviathan. *Subnautica*, Unknown Worlds Entertainment, 2018. Image credit: *Subnautica* Fandom Wiki.

The player, however, has the capacity to learn from the Precursors' past. They work with the SEL to develop a cure that is mutually beneficial for humans (in this case, the sole human alive on Planet 4546B), for the SEL's children and for the entire planet. Technology is the bridge between them, affording the player means to synthesise a cure using their crafting table, a cure borne from native plant-life found all over Planet 4546B in conjunction with biological matter from the SEL herself. However, the planet's fauna act as a technology of dispersion: the ever-present Peepers are able to carry the cure throughout the oceans, passing the cure on to all infected beings through a shimmering golden dispersal. Visually, the Kharaa Bacterium washes from the flora and fauna, particularly when Peepers are present. Envirofuturism is the cure after all: the Precursors were unwilling to embrace envirofuturism, were unwilling to engage in a mutually beneficial future. Players must choose this pathway to complete the game.

"We are different. But we go … together".

The game ends on a hopeful note laden with lingering questions, concerns and qualms. Upon curing the Kharaa Bacterium and building the escape rocket, players initiate the final launch sequence. Opportunities abound to leave a virtual time capsule behind for future players, with screenshots taken on this playthrough along with additional items to assist the future player in their own narrative.

There is an unspoken choice for the player as to whether they dismantle all of their structures and technology aside from the rocket platform, or leave it all as is, a remnant of their time on Planet 4546B. On the one hand, this is something of a moot point as far as video games are concerned: players can erase the game, effectively wiping their progress and any save state associated with the game. But, on the other, the player is presented with an implicit meaningful choice: clean up after themselves and leave the planet better than they left it, or, follow the path of the Precursors and leave their technology behind?

Upon reaching suborbital space and launching through the phasegate towards their home planet, the player is treated to one final telepathic vision from the SEL as she warmly imparts: "What is a wave without the ocean? A beginning without an end? They are different, but they go together. Now you go among the stars, and I fall among the sand. We are different. But we go … together." The SEL's poetic dialogue emphasises oneness with the player, a utopian call for the SEL's ideals to travel far beyond Planet 4546B and permeate the universe beyond. The player becomes a charge of the SEL, an envirofuturist.

Conversely, the last line of dialogue in *Subnautica* arrives just after this, after the credits roll: a foreboding reminder that the Alterra Corporation has no intention of leaving Planet 4546B alone, and in fact, that the player is chained by capitalism despite their best efforts. The screen fades to black, and the player hears a robotic, dispassionate voice state: "Welcome home to Alterra. Permission to land will be granted once you have settled your outstanding balance of one trillion credits." The implications are as damning as the SEL's last line is hopeful, yet after several dozen hours of gameplay and progressing through the story, neither should come as a surprise. The player not only is banned from landing, an inversion of Planet 4546B's quarantine

Envirofuturism

that protected the rest of the universe from the bacterium, but they are only permitted to land once they pay off their debt from harvesting minerals on the planet – which Alterra claims to own, just as the Precursors once claimed to own.

The subaquatic monster is anything but monstrous, a misunderstood mother looking after her children, themselves stewards of the oceanic planet. Likewise, the player's character returns to their "real" life back home on Alterra, and as the game ends, so too does the player return to reality. *Subnautica* challenges the player to pick up this envirofuturistic thread, "go among the stars" – or in the case of Earth, go among our primarily unexplored oceanic depths – and think critically about their position as a potential envirofuturist advocate in an otherwise capitalistic and colonial universe.

Bibliography

"Abdullah the Fisherman and Abudullah the Merman," in *One Thousand and One Nights*. Trans. Richard Francis Burton. Shammar ed. N. 547. Vol. 9 (1885–8), 165–88.

Adamowsky, Natascha, *The Mysterious Science of the Sea, 1775–1943* (Abingdon: Routledge, 2016).

Adams, Carol, *The Sexual Politics of Meat: A Feminist-Vegetarian Critical Theory* (Cambridge: Polity, 1990).

Agamben, Giorgio, *Stanzas: Word and Phantasm in Western Culture* (Minneapolis and London: Minneapolis University Press, 1993).

Aitken, Stuart C., "The Edge of the World: Embattled Leagues of Children and Seals Teeter on the Rim," in *Yearbook of the Association of Pacific Coast Geographers* 72 (2010), 12–32.

Alaimo, Stacy, "Introduction: Science Studies and the Blue Humanities," in *Configurations* 27/4 (2019), 429–32.

Alaimo, Stacy, "Violet-Black," in Jeffrey Jerome Cohen, ed., *Prismatic Ecologies: Ecotheory Beyond the Green* (Minneapolis: University of Minnesota Press, 2013), 233–51.

Alberti, Samuel J. M. M., *Morbid Curiosities: Medical Museums in Nineteenth-Century Britain*. (Oxford: Oxford University Press, 2011).

Alf Leilah w Leilah [One Thousand and One Nights], "Arous al-Bohoor" [The Maiden of the Seas], dir. Fahmy Abdel-Hameed (Cairo: Egyptian Television Productions, 1985).

American Cinematheque, "*Wolfwalkers*: Q&A with Tomm Moore and Ross Stewart, moderated by Guillermo del Toro," 14 February 2021. <https://youtu.be/DcLY VVwR_1A>. Accessed 5 May 2022.

Andersen, Hans Christian, "The Great Sea Serpent," in *The Complete Fairy Tales and Stories*, trans. Erik Christian Haugaard (New York: Anchor Books, 1983), 1006–14.

Andersen, Hans Christian, "The Little Mermaid" [1837], *Project Gutenberg*, N.D. <http://www.gutenberg.org/files/27200/27200-h/27200-h.htm#li_mermaid>. Accessed 6 June 2022.

Andersen, Hans Christian, "The Little Mermaid" [1837], trans. R. P. Keigwin, in Maria Tatar, ed., *The Classic Fairy Tales* (New York: W. W. Norton, 1999), 216–32.

Andersen, Hans Christian, *The Little Mermaid* (Great Britain: Hythloday Press, 2014).

Andersen, Hans Christian, "The Ugly Duckling" [1843], H.C. Andersen Centre, trans. Jean Hersholt 2019. <https://andersen.sdu.dk/vaerk/hersholt/TheUglyDuckling_e.html>. Accessed 1 October 2021.

Andrews, *Tamra, Dictionary of Nature Myths: Legends of the Earth, Sea, and Sky* (Oxford: Oxford University Press, 2000).

Anonymous, "Dugongo," Rabbitique, n.d.-a. <https://www.rabbitique.com/profile/pt/dugongo>. Accessed 6 June 2022.

Anonymous, "The 21st Century's 100 Greatest Films," *BBC Culture*, n.d.-b <https://www.bbc.com/culture/article/20160819-the-21st-centurys-100-greatest-films>. Accessed 13 June 2022.

Anonymous, *House Un-American Activities Committee*, at trumanlibrary.gov, n.d.-c <https://www.trumanlibrary.gov/education/presidential-inquiries/house-un-american-activities-committee>. Accessed 11 February 2022.

Ansell, Nicholas, "'Fantastic Beasts and Where to Find The(ir Wisdo)m': Behemoth and Leviathan in the Book of Job," in Koert van Bekkum, et al., eds, *Playing with Leviathan: Interpretation and Reception of Monsters from the Biblical World* (Leiden: Brill, 2017), 90–114.

Antón Sanchéz, Laura, "When Woman Teams Up with Monster: A Metafictional Discourse on Female Desire," in *Comparative Cinema* 8/15 (2020), 60–86.

Ashliman, D. L., "Melusina," *sites.pitt.edu*, 2015. <https://sites.pitt.edu/~dash/melusina.html>. Accessed 3 December 2021.

Atkinson, Michael, "Trouble Every Day: From *Amour Fou* to *Primal Scream*: Inside the Movie Madhouse of Andrzej Zulawski," in *Film Comment* 39/1 (2003), 38–41.

Atwood, Margaret, *Payback: Debt and the Shadow Side of Wealth* (Toronto: House of Anansi Press, 2008).

Austern, Linda Phyllis and Inna Naroditskaya, (eds), *Music of the Sirens* (Bloomington and Indianapolis: Indiana University Press, 2006).

Aydemir, Murat, *Images of Bliss: Ejaculation, Masculinity, Meaning* (Minneapolis: University of Minnesota Press, 2007).

Ayer Mata Duyong, dir. M. Amin (Kuala Lumpur: Cathay, 1964).

Baaqueel, Nuha Amad, *The Kaleidoscope of Gendered Memories in Ahlam Mosteghanemi's Novels* (Newcastle: Cambridge Scholars Press, 2019).

Bacchilega, Cristina and Marie Alohalani Brown, *The Penguin Book of Mermaids* [2015] (New York: Penguin Random House, 2019).

Back to the Black Lagoon: A Creature Chronicle, dir. David J. Skal (Los Angeles: Universal Pictures, 2002).

Badley, Linda, *Writing Horror and the Body: The Fiction of Stephen King, Clive Barker and Anne Rice.* (London: Greenwood Press, 1996).

Bain, Frederica, "The Tail of Melusine: Hybridity, Mutability, and the Accessible Other," in Misty Urban, Deva F. Kemmis, and Melissa Ridley Elmes, eds, *Melusine's Footprint: Tracing the Legacy of a Medieval Myth* (Leiden: Brill, 2017), 17–35.

Ballentine, Debra Scoggins, *The Conflict Myth and the Biblical Tradition* (New York: Oxford University Press, 2015).

Banner, L. W., "The Creature from the Black Lagoon: Marilyn Monroe and Whiteness," *Cinema Journal* 47/4 (Summer, 2008), 4–29.

Barratt-Peacock, Ruth, "The End of Words at the World's End: An Anthropocene Reading of David Malouf's 'The Crab Feast,'" in Gina Comos and Caroline Rosenthal, eds, *Anglophone Literature and Culture in the Anthropocene* (Berlin: Cambridge Scholars 2019), 176–95.

Barratt-Peacock, Ruth, *Concrete Horizons: Romantic Irony in the Poetry of David Malouf and Samuel Wagan Watson* (Bern: Peter Lang, 2020).

Bastine, Michael and Mason Winfield, *Iroquois Supernatural: Talking Animals and Medicine People* (Vermont: Bear & Company, 2011).

Bates, Daniel, "Could man soon be able to breathe underwater? Scientists eye possibility of merging human and algae DNA," *Mail Online*, 6 April 2011. <https://www.dailymail.co.uk/sciencetech/article-1374056/Could-man-breathe-underwater-Scientists-eye-possibility-merging-human-algae-DNA.html>. Accessed 3 June 2022.

Batto, Bernard F., *Slaying the Dragon: Mythmaking in the Biblical Tradition* (Atlanta: Westminster John Knox, 1992).

Beer, Gillian, *Open Fields: Science in Cultural Encounter* (Oxford: Oxford University Press, 1996).

Bellot, Gabrielle, "Dear Internet: The Little Mermaid also Happens to Be Queer Allegory." *Literary Hub*, 12 July 2019. Available at: <https://lithub.com/dear-internet-the-little-mermaid-also-happens-to-be-queer-allegory>. Accessed 1 October 2021.

Bennett, Emmett L. Jr., *The Pylos Tablets: Texts of the Inscriptions Found: PT II* (Princeton: Princeton University Press, 1955).

Bennett, George, "Death Mermaid, Eco-theater Fail to Sway Board on U.S. Sugar Land," *Palm Beach Post*, 12 March 2015. <https://www.palmbeachpost.com/article/20150312/NEWS/812065185>. Accessed 6 June 2022.

Bernstein, Robert and Ramona Fradon, "How Aquaman Got His Powers!," in Mort Weisinger, ed., *Adventure Comics #260* (New York: National Comics Publications, 1959), 17–24.

Bhabha, Homi, *The Location of Culture* [1994] (London and New York: Routledge, 2000).

Blackwood, Algernon, "The Willows,'" in S. T. Joshi, ed., *Ancient Sorceries and Other Weird Stories* (New York: Penguin, 2002), 17–62.

Blum, Hester, *The View from the Masthead: Maritime Imagination and Antebellum American Sea Narratives* (Chapel Hill: University of North Carolina Press, 2008).

Bly, Robert, *Iron John* (New York: Addison-Wesley, 1990).

Bolter, Jay David and Richard Grusin, *Remediation: Understanding New Media* (Cambridge: MIT Press, 1999).

Botting, Fred, *Limits of Horror: Technology, Bodies, and Horror* (Manchester: Manchester University Press, 2008).

Braham, Persephone, "Siren," in Jeffrey Andrew Weinstock, ed., *The Ashgate Encyclopedia of Literary and Cinematic Monsters* (Farnham: Ashgate, 2014), 516–19.

Braidotti, Rosi, *Nomadic Theory: The Portable Rosi Braidotti* (New York: Columbia University Press, 2011).

Braidotti, Rosi, *The Posthuman* (Cambridge: Polity, 2013).

Braidwood, Ella. "'Like moving through water while everyone is on land': The Writers Exploring Sexuality Through Sea Life," *The Guardian*, 3 February 2023. <https://www.theguardian.com/books/2023/feb/03/queer-writers-exploring-sexuality-and-gender-through-sea-life?CMP=Share_iOSApp_Other>. Accessed 21 February 2023.

Breen, J. L., *Letter from Joseph Breen to William Gordon, (2nd September 1953)*. Margaret Herrick Library Digital Collection. Motion Picture Association of America: Production Code Administration Records, 1953. Catalogue Record: http://catalog.oscars.org/vwebv/holdingsInfo?bibId=66279. Available at: <https://digitalcollections.oscars.org/digital/collection/p15759coll30/id/2794>. Accessed 5 June 2023.

Brooker, Will, *Batman Unmasked; Analysing a Cultural Icon* (London: Continuum, 2000).

Brown, William P., *Wisdom's Wonder: Character, Creation, and Crisis in the Bible's Wisdom Literature* (Grand Rapids: Eerdmans, 2014).

Bruford, Alan, "Trolls, Hillfolk, Finns, and Picts: The Identity of the Good Neighbors in Orkney and Shetland," in Peter Narvaez, ed., *The Good People: New Fairylore Essays* (Lexington: University Press of Kentucky, 1997), 116–41.

Brunson, Molly, *Russian Realisms. Literature and Painting, 1840–1890* (DeKalb: Northern Illinois University Press, 2016).

Buck, R. J., "Mycenaean Human Sacrifice," *Minos* 24 (1989), 131–7.

Bukatman, Scott, *Hellboy's World: Comics and Monsters on the Margins* (Berkeley: University of California Press, 2016).

Bull, Michael, *Sirens* (New York: Bloomsbury, 2020).

Burrows, Adam, *Time, Literature, and Cartography After the Spatial Turn* (New York: Palgrave Macmillan, 2016).

Burt, Jonathan, *Animals in Film* (London: Reaktion Books, 2002).

Callaghan, Gavin, *H. P. Lovecraft's Dark Acadia: The Satire, Symbology and Contradiction* (Jefferson: McFarland, 2013).

Callicott, J. Baird, *Earth's Insights: A Multicultural Survey of Ecological Ethics from the Mediterranean Basin to Australian Outback* (Berkeley: University of California Press, 1994).

Callier, Viviane, "Baby's Cells Can Manipulate Mom's Body for Decades," in *Smithsonian Magazine*, 2 September 2015. <https://www.smithsonianmag.com/science-nature/babys-cells-can-manipulate-moms-body-decades-180956493/>. Accessed 24 June 2022.

Campbell, Vincent, "Live Every Week Like It's Shark Week: Jaws and the Natural History Documentary," in I. Q. Hunter and Matt Melia, eds, *The Jaws Book: New Perspectives on the Summer Blockbuster* (New York: Bloomsbury Academic, 2020), 251–66.

Carbone, Marco Benoît, "Beauty and the Octopus: Close encounters with the other-than-human," in Jon Hackett and Seán Harrington, eds, *Beasts of the Deep: Sea Creatures and Popular Culture* (East Barnet: John Libbey, 2018).

Carrington, Richard, *Mermaids and Mastodons: A Book of Natural and Unnatural History* (London: Chatto and Windus, 1957).

Cashdan, Sheldon, *The Witch Must Die: The Hidden Meaning of Fairy Tales* (New York: Basic Books, 1999).

Cavallaro, Dani, *The Anime Art of Hayao Miyazaki* (Jefferson: McFarland, 2006).

Charles, Douglas M., "Communist and Homosexual: The FBI, Harry Hay, and the Secret Side of the Lavender Scare, 1943–1961," in *American Communist History* 11/1 (2012), 101–24.

Chemers, Michael M., "Le Freak, C'est Chic: The Twenty-First Century Freak Show as Theatre of Transgression," in *Modern Drama* 46/2 (2003), 285–304.

Christiansen, Reidar Thoralf, *Migratory Legends: List of Types with a Systematic Catalogue of the Norwegian Variants* (North Stratford: Ayer Co Pub, 1977).

Clare, Eli, *Exile and Pride: Disability, Queerness, and Liberation* (Durham: Duke University Press, 2015).

Clark, Cameron, "Grief, Ecocritical Negativity, and the Queer Anti-pastoral," in *New Review of Film and Television Studies* 17/2 (2019), 211–35.

Clarke, Timothy, *Ecocriticism on the Edge: The Anthropocene as a Threshold Concept* (London: Bloomsbury, 2015).

Clover, Carol J., *Men, Women and Chainsaws* (Princeton: Princeton University Press, 1992).

Cohen, Jeffrey Jerome, "Monster Culture: Seven Theses," in Jeffrey Jerome Cohen, ed., *Monster Theory: Reading Culture* (Minneapolis: University of Minnesota Press, 1996), 3–25.

Coney Island U. S. A., "The Mermaid Parade," *Coney Island U. S. A.*, 2022. <http://www.coneyisland.com/programs/mermaid-parade> Accessed 13 June 2022.

Cook, Erwin and Thomas Palaina, "New Perspectives on Pylian Cults Sacrifice and Society in the 'Odyssey,'" *Classical Studies*, 3 June 2016. <https://classicalstudies.org/sites/default/files/documents/abstracts/cook.pdf>. Accessed 28 April 2022.

Cook, W. James, *The Solossal P. T. Barnum Reader: Nothing Else Like It in the Universe – Phineas T. Barnum* (Urbana and Chicago: University of Illinois Press, 2005).

Cormick, Hanna, "Hanna Cormick, Performance Artist," *Art Space*, 2021. <https://arts accessaustralia.org/art-spaces/hanna-cormick/>. Accessed 23 September 2021.

Cornplanter, Jesse J., *Legends of the Longhouse* (Ontario: Irografts, 1986).

Coulter, Charles Russel and Patricia Turner, *Encyclopedia of Ancient Deities* (New York: Routledge, 2012).

Covill, Max., "How the Meg Made it to the Big Screen," *Film School Rejects*, 15 August 2018. <https://filmschoolrejects.com/the-meg-production-history/>. Accessed 11 August 2021.

Craig, Robert D., *Handbook of Polynesian Mythology* (Santa Barbara: ABC Clio, 2004).

Crane, Kylie, "Anthropocene and the End of the World: Apocalypse, Dystopia, and Other Disasters," in Gina Comos and Caroline Rosenthal, eds, *Anglophone Literature and Culture in the Anthropocene* (Berlin: Cambridge Scholars 2019), 158–75.

Crawl, dir. Alexandre Aja (Los Angeles: Paramount Pictures, 2019).

Creature from the Black Lagoon, dir. Jack Arnold (Los Angeles: Universal Pictures, 1954).

Creature Walks Among Us, The, dir. John Sherwood (Los Angeles: Universal Pictures, 1956).

Creed, Barbara, *The Monstrous-Feminine: Film, Feminism, Psychoanalysis* (Abingdon: Routledge, 1993).

Creed, Barbara, *The Monstrous Feminine: Film, Feminism, Psychoanalysis* [1997] (London: Routledge, 2015).

Cremin, Ciara, *The Future is Feminine: Capitalism and the Masculine Disorder* (London: Bloomsbury Academic, 2021).

Cronin, Brian, "How Aquaman's Final Issue Showed Up at Two Different Comic Companies," CBR.com, 12 June 2019. <https://www.cbr.com/aquaman-namor-eerie-steve-skeates-final-issue/>. Accessed 15 March 2021.

Cronin, Brian, "What Happened to Aquaman's 'Needs Water Every Hour or He'll Die' Limit?," CBR.com, 31 December 2018. <https://www.cbr.com/aquaman-hour-water-limit/>. Accessed 20 March 2021.

Darkside of Seoul Podcast, "Divine Monks and Shapeshifters: Korean Folktale Types," N.D. <https://www.darksideofseoul.com/divine-monks-shapeshifters-korean-folktale-types/>. Accessed 10 March 2021.

Davis, Lennard J., *Enforcing Normalcy: Disability, Deafness and the Body* (London: Verso, 1995).

de Bruyn, Ben, "The aristocracy of objects: shops, heirlooms and circulation narratives in Waugh, Fitzgerald and Lovecraft," in *Neohelicon* 42 (2015), 85–104.

Dekker, Jaap, "God and the Dragons in the Book of Isaiah," in Koert van Bekkum, et al., eds, *Playing with Leviathan: Interpretation and Reception of Monsters from the Biblical World* (Leiden: Brill, 2017), 21–39.

Demicheli, Dino, "Altar of the goddess Salacia from Trogir," *Opvscvla archaeologica* 31/1 (2007), 69–80.

Department of Veterans Affairs, "America's Wars: American War Deaths by Conflict," May 2021. <https://www.va.gov/opa/publications/factsheets/fs_americas_wars.pdf>. Accessed 6 June 2022.

Derry, Charles, *Dark Dreams 2.0: A Psychological History of the Modern Horror Film from the 1950s to the 21st Century* (Jefferson: McFarland & Company, 2009).

di Liddo, Annalisa, *Alan Moore: Comics as Performance, Fiction as Scalpel* (Jackson: University Press of Mississippi, 2009).

Dickens, Lyn, "Aquaman is the Mixed Race Movie We Didn't Know We Needed," medium.com, 23 January 2019. <https://medium.com/@lyn.dickens/aquaman-is-the-mixed-race-movie-we-didnt-know-we-needed-57841939feb6>. Accessed 8 March 2021.

Dikshitar, V. R. Ramachandra, *The Matsya Purana a Study* (Madras: University of Madras, 1935).

Dimmitt, Cornelia and J. A. B. van Buitenen, *Classical Hindu Mythology a Reader in the Sanskrit Puranas* (Philadelphia: Temple University Press, 1978).

Dinnerstein, Dorothy, *The Mermaid and the Minotaur: Sexual Arrangements and Human Malaise* (New York: Other Press, 1976).

Dirven, Lucinda, "The Author of De Dea Syria and his cultural heritage," *Numen* 44/2 (May 1997), 153–79.

Doak, Brian R., *Consider Leviathan: Narratives of Nature and the Self in Job* (Minneapolis: Fortress Press, 2014).

Doane, Mary Ann, *Femmes Fatales: Feminism, Film Theory, Psychoanalysis* (London: Routledge, 1991).

Doniger O'Flaherty, Wendy, *Hindu Myths* (New York: Penguin Random House, 1975).

Dring, M. J., *The Biology of Marine Plants* (Cambridge: Cambridge University Press, 2003).

Duffett, Mark and Jon Hackett, *Scary Monsters: Monstrosity, Masculinity and Popular Music* (New York: Bloomsbury, 2021).

Durbach, Nadja, *Spectacle of Deformity: Freak Shows and Modern British Culture* (Berkeley: University of California Press, 2009).

Duyong Aridinata, dir. Adam Hamid (Kula Lumpur: Zeel Productions, 2010).

Duyung, dir. Abdul Razak Mohaideen (Kuala Lumpur: KRU Studios, 2008).

Dyett, Jordan and Thomas Cassidy, "Overpopulation Discourse: Patriarchy, Racism, and the Specter of Ecofascism," in *Perspectives on Global Development and Technology* 18/1–2 (2019), 205–24.

Ebbatson, Roger, *Landscapes of Eternal Return: From Tennyson to Hardy* (Cham: Palgrave Macmillan, 2016).

Egan, Kieran, "Fantasy and Reality in Children's Stories," *Simon Fraser University: Kieran Egan.* <https://www.sfu.ca/~egan/FantasyReality.html>. Accessed 3 February 2021.

Eisenstein, Sergei, *On Disney.* Ed. Jay Leyda. Trans. Alan Upchurch (London: Seagull Books, 2017).

Elliott, Brian, *The Landscape of Australian Poetry* (Melbourne: Cheshire Publishing, 1967).

Ellis, Heather, "Knowledge, Character and Professionalisation in Nineteenth-Century British Science," in *History of Education* 43/6 (2014), 777–92.

Emerson, Caryl, "To What End Rusalka? Pushkin's Folk Tragedy and Dargomyzhskii's Opera," in *The Slavonic and East European Review* 97/1 (2019), 169–200.

Encyclopaedia Britannica, The [9th edn]: vol. XXI (Edinburgh: Adam and Charles Black, 1886).

Evolution, dir. Lucile Hadžihalilović (Paris: Potemkine Films, 2015).

Eye & Mermaid [*Houreya wa Ein*], dir. Shahad Ameen (Doha: The Doha Film Institute, 2013).

Fadhlaoui, Anis, Mohamed Khrouf, Soumaya Gaigi, Fethi Zhioua and Anis Chaker, "The Sirenomelia Sequence: A Case History," in *Clinical Medicine Insights: Case Reports*, (January 2010).

Falconer, Dalia, "The Opposite of Glamour," *Sydney Review of Books.com*, 28 July 2017. <https://sydneyreviewofbooks.com/essay/the-opposite-of-glamour-delia-falco ner/>. Accessed 26 October 2021.

Farrimond, Katherine, *The Contemporary Femme Fatale: Gender, Genre and the American Cinema* (New York: Routledge, 2018).

Fehrenbach, T. R., *This Kind of War: A Study in Unpreparedness* (New York: Macmillan, 1963).

FishBase, "A Globar Information System of Fishes," *FishBase*, 2021. <https://www.fishb ase.de/home.htm>. Accessed 2 December 2021.

Fiske, John, Bob Hodge and Graeme Turner, *Myths of OZ* (Sydney: Allen and Unwin, 1987).

Fitzsimmons, Phil, "The Mything Link: The Feminine Voice in the Shifting Australian National Myth," in Colleen Harris and Valerie Estelle Frankel, eds, *Woman Versed in Myth* (Jefferson: McFarland, 2016), 106–13.

Fletcher, Lisa and Daniel Crane, *Island Genre, Genre Island: Conceptualisation and Representation in Popular Fiction* (Washington, DC: Rowman and Littlefield, 2017).

Fraser, Lucy, "Reading and Retelling Girls across Cultures: Mermaid Tales in Japanese and English," *Japan Forum* 26/2 (2014), 246–64.

Frasl, Beatrice, "Bright Young Women, Sick of Swimmin' Ready to … Consume? The Construction of Postfeminist Feminity in Disney's *The Little Mermaid*," *European Journal of Women's Studies* 25/3 (2018), 341–54.

Gadallah, Moustafa, *Isis: The Divine Female* (Greensboro: Tehuti Research Foundation, 2016).

Gaiman, Neil and P. Craig Russell, *Only the End of the World Again* (Portland: Oni Press, 2000).

Gauthier, Michael, "Enviro-Futurism," *Environmental League*, 31 May 2018. <https://www.theenvironmentalleague.com/post/enviro-futurism>. Accessed 24 June 2018.

Gerber, David A., "The 'Careers' of People Exhibited in Freak Shows: The Problem of Volition and Valorization," in Rosemarie Garland-Thompson, ed., *Freakery: Cultural Spectacles of the Extraordinary Body* (New York: New York University Press, 1996), 38–54.

Gilbert, Asha C., "A Man on a California Beach Thought He Saw a Football. It Was a Rare Deep Sea Monster," *USA Today*, 30 November 2021. <https://www.usatoday.com/story/news/nation/2021/11/30/rare-pacific-footballfish-california-beach/8804818002/>. Accessed 1 December 2021.

Gilbert, James, *Men in the Middle: Searching for Masculinity in the 1950s* (Chicago: The University of Chicago Press, 2005).

Goddard, Michael, "Beyond Polish Moral Realism: The subversive Cinema of Andrzej Żuławski," in Eva Mazierska and Michael Goddard, eds, *Polish Cinema in a Transnational Context* (Rochester: University of Rochester Press, 2014), 236–57.

Gogol, Nikolai, *Evenings Near the Village of Dikanka* [1831]. Ed. Ovid Gorchakov (Moscow: Foreign Languages Publishing House, 1957).

Gogol, Nikolai, *Mirgorod. Being a Continuation of Evenings in a Village near Dikanka* [1835]. Trans. A. Kanevsky (Moscow: Foreign Languages Publishing House, 1958).

Goodlow, Ollie G., Roslind McCoy Sibley, Bonnie G. Allen, William Saa Kamanda, Allyce C. Gullattee and William C. Rayfield, "Sirenomelia: Mermaid Syndrome," in *Journal of the National Medical Association*, 80/3 (1988), 343–6.

Govindrahan, Radhika, *Animal Intimacies: Interspecies Relatedness in India's Central Himalayas* (Chicago: University of Chicago Press, 2018).

Grafius, Brandon R., *Reading the Bible with Horror* (Lanham: Lexington Books/Fortress Academic, 2019).

Grant, Mira, *Into the Drowning Deep* (New York: Orbit, 2017).

Gregersdotter, Katarina and Nicklas Hållén, "Anthropomorphism and the Representation of Animals as Adversaries," in Katarina Gregersdotter, Johan Höglund and Nicklas Hållén, eds, *Animal Horror Cinema: Genre, History and Criticism* (London: Palgrave Macmillan, 2015), 206–23.

Gregersdotter, Katarina, Nicklas Hållén and Johan Höglund, "Introduction," in Katarina Gregersdotter, Johan Höglund and Nicklas Hållén, eds, *Animal Horror Cinema: Genre, History and Criticism* (London: Palgrave Macmillan, 2015), 1–18.

Gregersdotter, Katarina, Nicklas Hållén and Johan Höglund, "A History of Animal Horror Cinema," in Katarina Gregersdotter, Johan Höglund and Nicklas Hållén, eds, *Animal Horror Cinema: Genre, History and Criticism* (London: Palgrave Macmillan, 2015a), 18–36.

Griffin, Sean, *Tinker Belles and Evil Queens: The Walt Disney Company from the Inside Out* (New York: New York University Press, 2000).

Grimaud, Jessica, "Where Is Moana from? Discover the Real Heritage of Disney's Latest Princess," *FamilySearch*, 8 January 2021. <https://www.familysearch.org/blog/en/where-is-moana-from/>. Accessed 23 June 2021.

Grusin, Richard, (ed.), *Anthropocene Feminism* (Minnesota: University of Minnesota Press, 2017).

Guest Contributor, "Acclaimed Mermaid Delivers Strong Message to Chicken of the Sea," *Eco Watch*, 2 October 2015. <https://www.ecowatch.com/acclaimed-merm aid-delivers-strong-message-to-chicken-of-the-sea-1882115105.html>. Accessed 6 June 2022.

Gulizio, Joann, "Hermes and e-ma-a2: The Continuity of His Cult from the Bronze Age to the Historical Period," *Živa Antika* 50 (2000), 105–16.

Gumilev, Nikolaj Stepanovič, *Polnoe sobranie sočinenij v 10 tomah. Tom 1.* Stihotvorenija. Poèmy (1902–1910) (Moskva: Voskresenie, [1905] 1998).

Gutierrez, Anna Katrina, "Metamorphosis: The Emergence of Glocal Subjectivities in the Blend of Global, Local, East and West," in John Stephens, ed., *Subjectivity in Asian Children's Literature and Film: Global Theories and Implications* (London: Taylor & Francis, 2012), 19–42.

Hallerton, Sheila, "Atlantis/Lyonesse: The Plains of Imagination," *Shima* 10/2 (2016), 112–17.

Hamilton, Jennifer., "Cold Desire: Snow, Ice and Hans Christian Andersen," in A. Uhlmann, H. Groth, P. Sheehan and S. McClaren, eds, *Literature and Sensation* (Sydney: Cambridge Scholar's Press, 2009), 244–54.

Handmaiden, The, dir. Park Chan-wook (Seoul: CJ Entertainment, 2016).

Haraway, Donna, *The Companion Species Manifesto: Dogs, People, and Significant Otherness* (Chicago: Prickly Paradigm Press, 2012).

Hardin, Garret, "Interview with The Social Contract (21 June 1997)," *The Garret Hardin Society*, 9 June 2003. <https://www.garretthardinsociety.org/gh/gh_straub_interv iew.html>. Accessed 10 September 2020.

Harris, Jason Marc, "Perilous Shores: The Unfathomable Supernaturalism of Water in 19th-Century Scottish Folklore," *Mythlore* 28 (2009), 5–25.

Hauser, Brian R., "Weird Cinema and the Aesthetics of Dread," in Sean Moreland, ed., *New Directions in Supernatural Horror Literature: The Critical Influence of H. P. Lovecraft* (London: Palgrave Macmillan, 2018), 235–52.

Hayward, Philip, "From Dugongs to *sinetrons*: Syncretic Mermaids in Indonesian Culture," in Philip Hayward, ed., *Scaled for Success: The Internationalisation of the Mermaid* (Eastleigh: John Libbey and Indiana University Press, 2018b), 89–106.

Hayward, Philip, "Japan: The Mermaidisation of the *ningyo* and Related Folkloric Figures," in Philip Hayward, ed., *Scaled for Success: The Internationalisation of the Mermaid* (Eastleigh: John Libbey and Indiana University Press, 2018a), 51–68.

Hayward, Philip, "Mermaid Horror," *University of Technology Sydney* seminar paper (2016).

Hayward, Philip, *Making a Splash: Mermaids (and Mermen) in 20th and early 21st Century Audiovisual Media* (Eastleigh: John Libbey and Indiana University Press, 2017).

Hayward, Philip and Lisa Milner, "Shoreline Revels; Perversity, Polyvalence and Exhibitionism at Coney Island's Mermaid Parade," in Philip Hayward, ed., *Scaled for Success: The Internationalisation of the Mermaid (Eastleigh: John Libbey and University of Indiana Press, 2018), 196–209.*

Heaney, Seamus, "Introduction," in *People of the Sea*, by David Thomson (Edinburgh: Canongate Books, 2017).

Hellboy 2: The Golden Army, dir. Guillermo del Toro (Los Angeles: Universal Pictures, 2008).

Hellboy, dir. Guillermo del Toro (Los Angeles: Revolution Studios, 2004).

Helmreich, Stefan, *Alien Ocean: Anthropological Voyages in Microbial Seas* (Berkeley: University of California Press, 2009). <https://doi.org/10.1525/9780520942608>

Henault, Marielle Chartier, "How to Become a Professional Mermaid," *Aquamermaid*, 7 May 2020. <https://aquamermaid.com/blogs/news/how-to-become-a-professio nal-mermaid>. Accessed 6 June 2022.

Hewitt, John Napoleon Brinton and Jeremiah Curtin, *Seneca Fiction, Legends, and Myths* (Charleston: Nabu, 2010).

Hill, Kent. "Steve, The MEG & I: 20 Years in the Making (Part 2) by Kent Hill," *Podcasting Them Softly*, 28 August 2018. <https://podcastingthemsoftly.com/2018/ 08/28/steve-the-meg-i-20-years-in-the-making-part-2-by-kent-hill/>. Accessed 28 April 2022.

Hilton, Alison, *Russian Folk Art* (Bloomington and Indianapolis: Indiana University Press, 1995).

Holmes, John and Sharon Ruston, "Introduction: Literature and Science in the Nineteenth Century," in John Holmes and Sharon Ruston, eds, *The Routledge Research Companion to Nineteenth-Century British Literature and Science* (Abingdon: Routledge, 2017), 1–16.

Holmes, John, "Natural History, Evolution and Ecology," in John Holmes and Sharon Ruston, eds, *The Routledge Research Companion to Nineteenth-Century British Literature and Science* (Abingdon: Routledge, 2017), 331–56.

Homer, "Book XII," in *Odyssey*, trans. Robert Fagles (London: Penguin, 2006), 48–57.

Honeycutt, J. L., et al. "Warmer Waters Masculinize Wild Populations of a Fish with Temperature-Dependent Sex Determination," *Scientific Reports* 9 (2019), 1–13.

Host, The, dir. Bong Joon-ho (Seoul: Showbox Entertainment, 2006).

Hughes, Dennis D. *Human Sacrifice in Ancient Greece* (London and New York: Routledge, 1991).

Humanoids from the Deep, dir. Barbara Peeters (Los Angeles: New World Pictures, 1980).

Hunt, Elle, "An Octopus 'Love Story' on Netflix Has Caused Thoughts to Run Wild. Why?" *The Guardian*, 28 September 2020. <https://www.theguardian.com/commentisfree/2020/sep/24/octopus-love-story-netflix>. Accessed 29 May 2021.

Hunter, I. Q., *Cult Film as a Guide to Life. Fandom, Adaptation and Identity* (New York: Bloomsbury, 2016).

Illustrated London News, The, 28 October 1848.

Irigaray, Luce, *This Sex Which Is Not One* [1977], trans. by Catherine Porter with Carolyn Burke (Ithaca: Cornell University Press, 1985).

Irvine, Scott, *Ishtar and Ereshkigal: The Daughters of Sin* (Hampshire: Moon Books, 2020).

Ivanits, Linda J., *Russian Folk Belief* (Armonk, New York and London: M. E. Sharpe, 1989).

Jackson, Neil., "Ben Gardner's Head is Missing: Notes on *Jaws: The Sharksploitation Edit*," in I. Q. Hunter and Matt Melia, eds, *The Jaws Book: New Perspectives on the Classic Summer Blockbuster* (New York: Bloomsbury, 2020).

James, Megan, *Innsmouth* (Richmond: Sink/Swim Press, 2016–18).

Jarvis, Shawn C., "Mermaid", in D. Haasse, ed., *The Greenwood Encyclopedia of Folktales and Fairy Tales, Volume 1* (Greenwood Press, 2008), 619–22.

Johns, Geoff, Reis, Ivan and Joe Prado, *Aquaman Volume 1: The Trench* (Burbank: DC Comics, 2012).

Johnson, Katherine, *Sexuality: A Pyschosocial Manifesto* (Cambridge: Polity Press, 2015).

Jøn, Allan Asbjørn, "Dugongs and Mermaids, Selkies and Seals," in *Australian Folklore: A Yearly Journal of Folklore Studies* 13 (1998), 94–8.

Joshi, K. L., (ed.), *Matsya Purana* (Delhi: Parimal Publications, 2020).

Joyce, Thomas Andrew, "A Nation of Employees: The Rise of Corporations and the Perceived Crisis of Masculinity in the 1950s," in *The Graduate History Review*, 3/1 (2011), 24–48.

Kan Yama Kan [Once Upon a Time], "Arous al-Bahr" [The Mermaid], Season I, dir. Bassam al-Molla (Damascus: Sheikhani for Creative Visions, 1992).

Karpova, Tat'jana, *Ivan Kramskoj* (Moskva: Belyj gorod, 2000).

Keetley, Dawn, "Tentacular Ecohorror and the Agency of Trees in Algernon Blackwood's 'The Man Whom the Trees Loved' and Lorcan Finnegan's *Without Name*," in Christy Tidwell and Carter Soles, eds, *Fear and Nature: Ecohorror Studies in the Anthropocene* (Pennsylvania: Penn State University Press, 2021), 23–41.

Keller, Catherine, *The Face of the Deep: A Theology of Becoming* (London: Routledge, 2003).

Keller, James R., *V for Vendetta as Cultural Pastiche: A Critical Study of the Graphic Novel and Film* (London: McFarland, 2008).

Kiernan, Caitlín, "The Mermaid of the Concrete Ocean," in Paula Guran, ed., *Mermaids and Other Mysteries of the Deep* (Germantown: Prime Books, 2015).

Kim, Kyung Hyun, *Virtual Hallyu: Korean Cinema of the Global Era* (Durham and London: Duke University Press, 2011).

Kimmel, Michael S., *Manhood in America: A Cultural History* [Second Edition] (New York: Oxford University Press, 2006).

King James V, *The Bible: Authorized King James Version*. Ed. Robert P. Carroll and Stephen Prickett (Oxford: Oxford University Press, 2008).

Kingsnorth, Paul, "Dark Ecology," *Orion Magazine*, 31 December 2012. <https://orionmagazine.org/article/dark-ecology/>. Accessed 28 November 2021.

Kipling, Rudyard, *A Song of the English* (New York: Doubleday, Page, 1909).

Kon, Satoshi, *Tropic of the Sea* (New York: Vertical, 2013).

Kononenko, R., *Ivan Nikolaevič Kramskoj, 1837–1887* (Moskva: Direkt-media, 2009).

Korpel, Marjo and Johannes de Moor, "The Leviathan in the Ancient Near East," in Koert van Bekkum, et al., eds, *Playing with Leviathan: Interpretation and Reception of Monsters from the Biblical World* (Leiden: Brill, 2017), 3–18.

Kramskoj, Ivan Nikolaevič, *Perepiska I. N. Kramskogo. Tekst I. N. Kramskoj i P. M. Tret'jakov. (1869–1887)* (Moskva: Iskusstvo, 1953).

Krishna, Nanditha et al., *The Book of Avatars and Divinities* (Haryana: Penguin Random House, 2018).

Kristeva, Julia, *Powern of Horror* (New York: Columbia University Press, 1982).

Kurlander, Eric, *Hitler's Monsters: A Supernatural History of the Third Reich* (New Haven and London: Yale University Press, 2017).

Kuročkina, Tat'jana, *Ivan Nikoiaevič Kramskoj* (Leningrad: Hudožnik RSFSR, 1989).

Kuznetski, Julia and Stacy Alaimo, "Transcorporeality: An Interview with Stacy Alaimo," *Ecozon@*, 2 November 2020. <https://DOI.ORG/10.37536/ECOZONA.2020.11.2.3478>. Accessed 24 November 2021.

Lacan, Jaques, *Book XX Encore 1972–1973: On Feminine Sexuality: The Limits of Love and Knowledge* (New York and London: W. W. Norton, 1998).

Langan, John, *The Fisherman* (Petaluma: Word Horde, 2016).

Launier, Kimberley. "Splash! Real-Life Mermaids Seek Beauty of Deep," *ABC News*, 28 May 2010. <https://abcnews.go.com/2020/real-life-mermaids-superhumans/story?id=10771939>. Accessed 28 April 2022.

Lee, Henry, *Sea Monsters Unmasked* (London: William Clowes and Sons, 1883).

Lermontov, Mihail Jur'evič, *Polnoe sobranie sočinenij M. Ju. Lermontova v' dvuh' tomah. Tom' I* (Moskva: Izdanie Tovariśestva M. G. Vol'f, 1896).

Levenson, Jon D., *Creation and the Persistence of Evil: The Jewish Drama of Divine Omnipotence* (Princeton: Princeton University Press, 1988).

Linkota, Pentti, *Can Life Prevail? A Radical Approach to the Environmental Crisis* (Budapest: Arketos Media Limited, 2011).

Little Mermaid, The, dir. Ron Clements and John Musker (Los Angeles: Buana Vista Pictures, 1989).

Lovecraft, H. P., *Supernatural Horror in Literature* (New York: Dover Publications, 1973).

Lovecraft, H. P., *The Complete Fiction of H. P. Lovecraft* (New York: Chartwell Books, 2015).

Loxley, Diana, *Problematic Shores: The Literature of Islands* (London: Palgrave Macmillan, 1990).

Luomala, Katharine, *Voices on the Wind: Polynesian Myths and Chants. Revised Edition* (Honolulu: Bishop Museum Press, 1986).

Lure, The [Córki dancingu], dir. Agnieszka Smoczyńska (Warsaw: Kino Świat, 2015).

Lyons, Sherrie Lynne, *Species, Serpents, Spirits, and Skulls: Science at the Margins in the Victorian Age* (New York: SUNY Press, 2009).

Macumber, Heather, "The Threat of Empire: Monstrous Hybridity in Revelation 13," *Biblical Interpretation* 27/1 (2019), 107–29.

Macumber, Heather, *Monstrous Visions: Hybridity and Liminality in John's Apocalypse* (Lanham: Lexington Books/Fortress Academic, 2021).

Malouf, David, *Poems 1959–89* (Queensland: University of Queensland Press, 1992).

Malouf, David, *Revolving Days: Selected Poems* (Queensland: University of Queensland Press, 2008).

Mani, Vettam, *Puranic Encyclopedia* (Delhi: Motilal Banarasaidass, 1975).

Marz, Ron and Ivan Rodriguez, *The Shadow Over Innsmouth* (Mt Laurel: Dynamite Entertainment, 2014).

Marzouk, Safwat, *Egypt as a Monster in the Book of Ezekiel* (Tübingen: Mohr Siebeck, 2015).

Mazurek, Marcin, *A Sense of the Apocalypse: Technology, Textuality, Identity* (New York: Peter Lang, 2014).

McCance, Dawne, *Critical Animal Studies: An Introduction* (Albany: State University of New York Press, 2013).

McCarrick, Michael, "The American Version of *Spirited Away* Makes a MAJOR Change to the Ending," CBR.com., 20 July 2021. <https://www.cbr.com/spirited-away-dis ney-change-ending/>. Accessed 19 June 2022.

McEntire, Nancy Cassell, "Supernatural Beings in the Far North: Folkore, Folk Belief, and the Selkie," *Scottish Studies* 35 (2010), 120–43.

McFarlane, Brian, "Reading Film and Literature," in Deborah Cartmell and Imeda Whelehan, eds, *The Cambridge Companion to Literature on Screen* (Cambridge: Cambridge University Press, 2007), 15–28.

Melissa, Mermaid, "Professional Mermaid Performer 'Real-life Mermaid' Mermaid for Hire," *Mermaid Melissa*, 2022. <http://www.mermaidmelissa.com/>. Accessed 13 June 2022.

Menon, Ramesh (trans.), *Bhagavata Purana* (Kindle Edition, 2012).

Meyers, Robert W., "The Little Mermaid: Hans Christian Andersen's Feminine Identification," *Journal of Applied Pyschoanalytic Studies* 3/2 (2001), 149–59.

Michael Rostovtzeff, "Hadad and Atargatis at Palmyra," *American Journal of Archeology* 37 (January 1933), 58–63.

Michelinie, David and Jim Aparao, "Dark Destiny, Deadly Dreams," in Joe Orlando and Paul Levitz, eds, *Adventure Comics #452* (New York: DC Comics, 1977), 2–19.

Mignola, Mike and John Arcudi, *B.P.R.D.: The Universal Machine #4* (Milwaukie: Dark Horse Comics, 2007).

Mignola, Mike, *B.P.R.D.: The Dead #5* (Milwaukie: Dark Horse Comics, 2005).

Mignola, Mike, et al., *Abe Sapien Volume 3: Dark and Terrible and the New Race of Man* (Milwaukie: Dark Horse Comics, 2013).

Mignola, Mike, *Hellboy* (Milwaukie: Dark Horse Comics, 1993–present).

Mignola, Mike, *Hellboy: Conqueror Worm #4* (Milwaukie: Dark Horse Comics, 2001).

Mignola, Mike, *Hellboy: Seed of Destruction* (Milwaukie: Dark Horse Comics, 1994).

Miller II, Robert D., *The Dragon, the Mountain, and the Nations: An Old Testament Myth, Its Origins, and Its Afterlives* (University Park: Eisenbrauns, 2018).

Mitchell, D. T. and S. L. Snyder, *Narrative Prosthesis: Disability and the Dependencies of Discourse* (Ann Arbor: University of Michigan Press, 2001).

Mitchell, Lauren A., "Erotic Surgery: J. G. Ballard's *Crash*, Octavia Butler's 'Bloodchild,' and the Visual Legacy of the Medical Museum," in *Configurations* (2022).

Miyazaki, Hayao, *The Art of Spirited Away* (San Francisco: VIZ Media LLC, 2012).

Moana, dir. Ron Clement, John Musker, Don Hall and Chris Williams (Burbank: Walt Disney Studios Motion Pictures, 2016).

Mobley, Gregory, *The Return of the Chaos Monsters – and Other Backstories of the Bible* (Grand Rapids: Eerdmans, 2012).

Monster Squad, The, dir. Fred Dekker (Los Angeles: Taft Entertainment/Home Box Office, 1987).

Monterey Bay Aquarium, "Deep Sea Anglerfish," montereybayaquarium.org, 2022. <https://www.montereybayaquarium.org/animals/animals-a-to-z/deep-sea-anglerfish>. Accessed 1 December 2021.

Moore, Alan and Jacen Burrows, *Neonomicon* (Rantoul: Avatar Press, 2010–11).

Moore, Alan and Jacen Burrows, *Providence* (Rantoul: Avatar Press, 2015–17).

Moore, Alan and Jacen Burrows, *The Courtyard* (Rantoul: Avatar Press, 1994).

Moore, Tomm, *Song of the Sea Artbook* (Kilkenny: Cartoon Saloon, 2016).

Moreland, Sean, "The Birth of Cosmic Horror from the S(ub)lime of Lucretius," in Sean Moreland, ed., *New Directions in Supernatural Horror Literature: The Critical Influence of H. P. Lovecraft* (London: Palgrave Macmillan, 2018), 13–42.

Morris, Jon, *The League of Regrettable Superheroes* (Philadelphia: Quirk Books, 2015).

Morrison, Grant, *Supergods: What masked Vigilantes, Miraculous Mutants, and a Sun God from Smallville can Teach Us About Being Human* (New York: Spiegel & Grau, 2011).

Mortensen, Finn Hauberg, "The Little Mermaid: Icon and Disneyfication," *Scandinavian Studies* 80/4 (2008), 437–53.

Motamayor, Rafael, "The Meg: It's Got Shark-Punching, but It's Also Got a Message about the Environment. Yes, Really," *The Guardian*, 16 August 2018. <https://www.theguardian.com/film/2018/aug/16/the-meg-shark-action-film-message-eco logical-destruction>. Accessed 11 August 2021.

Murphy, J. J., G. C. Fraser and G. K. Blair, "Sirenomelia: Case of the Surviving Mermaid," in *Journal of Pediatric Surgery* 27/10 (1992).

Murrow, Edward R., "A Report on Senator Joe McCarthy," *See It Now* (transcript), 9 March, 1954. <http://www.plosin.com/beatbegins/archive/Murrow540309. htm>. Accessed 11 February 2022.

My Octopus Teacher, dir. Pippa Ehrlich and James Reed (Los Gatos: Netflix, 2020).

Mylonopoulos, I., "Poseidon," in *The Encyclopedia of Ancient History* (Hoboken: Wiley Blackwell, 2012).

Nabokov-Sirin, Vladymir, "Rusalka. Zaključitel'naja scena k puškinskoj Rusalke," in *Novyj Žurnal* 2 (1942), 181–4.

Napier, Susan, *Miyazakiworld: A Life in Art* (New Haven: Yale University Press, 2018).

Naroditskaya, Ina, "Russian Rusalkas and Nationalism: Water, Power, and Women," in Linda Phyllis Austern and Inna Naroditskaya, eds, *Music of the Sirens* (Bloomington and Indianapolis: Indiana University Press, 2006), 216–49.

National Weather Service, "Layers of the Ocean." *Internet Archive*, 7 February 2017. <http://www.srh.noaa.gov/jetstream/ocean/layers_ocean.html>. Accessed 1 December 2021.

Neill, Sam, "I'm Very Sorry to Hear This. Zulawski Was a Brilliant Maniac. It Nearly Killed Me, but Work with Him Was Great," Twitter, 18 February 2016. <https://twitter.com/TwoPaddocks/status/700178368847679489 >. Accessed 29 May 2021.

Neimanis, Astrida, *Bodies of Water: Posthuman Feminist Phenomenology* (London: Bloomsbury Publishing, 2017).

Neumann, Erich, *The Great Mother: An Analysis of the Archetype* (Princeton: Bollingen Foundation, 1963).

Neumann, Erich, *The Great Mother: An Analysis of the Archetype* (Princeton: Princeton University Press, 2015).

Newitz, Annalee, "The Undead: A Haunted Whiteness," in Jeffrey Andrew Weinstock, ed., *The Monster Theory Reader* (Minneapolis: University of Minnesota Press, 2020), 241–71.

Newman, Dean., "Meg Facts: Jurassic Shark," *The Daily Jaws*, 7 August 2018. <https://thedailyjaws.com/blog/2018/8/7/the-meg-files-jurassic-shark-facts>. Accessed 11 August 2021.

Newman, Kim, *Nightmare Movies: Horror on Screen Since the 1960s* [1988] (London: Bloomsbury, 2001).

Newsom, Carol A., *The Book of Job: A Contest of Moral Imaginations* (New York: Oxford University Press, 2003).

No Comment TV, "Making a splash at Malaysia's mermaid school," Euronews, 6 March 2019. <https://www.youtube.com/watch?v=Uh2jrvzl4zU>. Accessed 6 June 2022.

Nordin, Irene Gilsenan, "'Between the Dark Shore and the Light': The Exilic Subject in Eiléan Ní Chuilleanáin's The Second Voyage," in Michael Boss, Irene Gilsenan Nordin and Britta Gilden, eds, *Remapping Exile: Realities and Metaphors in Irish Literature and History* (Aarhus: Aarhus University Press, 2006), 178–94.

Noson, Katie, "That Hateful Tail: The Sirena as Figure for Disability in Italian Literature and Beyond," in *California Italian Studies* 6/1 (2016).

O'Neill, Junella Pusbach, *The Great New England Sea Serpent: An Account of Unknown Creatures Sighted by Many Respectable Persons Between 1683 and the Present Day* (New York: Paraview, 2003).

O'Regan, Cyril, Gnostic Apocalypse: *Jacob Boehme's Haunted Narrative* (Albany: State University of New York Press, 2002).

O'Regan, Cyril, *Gnostic Apocalypse: Jacob Boehme's Haunted Narrative* (New York: State University of New York Press, 2012).

O'Sullivan, Lisa and Julie Anderson, "Histories of Disability and Medicine: Reconciling Historical Narratives and Contemporary Values," in Richard Sandell, ed., *Re-Presenting Disability: Activism and Agency in the Museum* (London: Taylor & Francis, 2010), 143–54.

Och, Dana, "The Missing Dead of the Great Hunger: Metaphor and Palimpsest in Irish Film," in Anastasia Ulanowicz and Manisha Basu, eds, *The Aesthetics and Politics of Global Hunger* (London: Palgrave Macmillan, 2017), 177–204.

Oldboy, dir. Park Chan-wook (Seoul: Show East, 2003).

Olsen, Barbara A., *Women in Mycenaean Greece: The Linear B Tablets from Pylos and Knossos* (London and New York: Routledge, 2014).

Op de Beeck, Nathalie, "Anima and Animé: Environmental Perspectives and New Frontiers in *Princess Mononoke* and *Spirited Away*," in Mark I. West, ed., *The Japanification of Children's Popular Culture: From Godzilla to Miyazaki* (Lanham: Scarecrow Press, 2009), 267–84.

Orlova, E., *Ivan Nikoiaevič Kramskoj* (Moskva: Fanki Ink, 2014).

Othman, Hussain, "Conceptual Understanding of Myths and Legends in Malay History," in *Sari* 26 (2008), 91–110.

Otto, Rudolph, *The Idea of the Holy*. Trans. John W. Harvey (New York: Oxford, 1950).

Oudemans, A. C., *The Great Sea-Serpent* (Leiden: E. J. Brill, 1892).

Packham, Jimmy and David Punter, "Oceanic Studies and the Gothic Deep," *Gothic Studies* 19/2 (2017), 16–29.

Parikka, Jussi, *A Geology of Media* (Minneapolis: University of Minnesota Press, 2015).

Paxton, C. G. M. and D. Naish, "Did Nineteenth Century Marine Vertebrate Fossil Discoveries Influence Sea Serpent Reports?," in *Earth Sciences History* 38/1 (2019), 16–27.

Perrault, Charles, "Little Red Hood," trans. Stanley Appelbaum, in Stanley Appelbaum, ed., *The Complete Fairy Tales in Verse and Prose: A Dual-Language Book* (New York: Dover Publications, 2002), 132–3.

Petkova, Savina, "Notebook Primer: Mermaid Cinema," *Mubi*, 15 April 2021. <https://mubi.com/notebook/posts/notebook-primer-mermaid-cinema>. Accessed 17 September 2021.

Piatti-Farnell, Lorna, *Consuming Gothic: Food and Horror in Film* (London: Palgrave Macmillan, 2017).

Ponyo, dir. Hayao Miyazaki (Tokyo: Studio Ghibli, 2008).

Possession, dir. Andrzej Żuławski (Neuilly-sur-Seine: Gaumont, 1981).

Priestley, Andrew, "Sea Creatures Protest against Marine Sanctuary Zones Being Opened for Fishing," Manly Daily November 6, 2013. <https://www.dailytelegraph.com.au/newslocal/northern-beaches/sea-creatures-protest-against-marine-sanctuary-zones-being-opened-for-fishing/news-story/4a3d31829619e04bb6d557249759c4a4>. Accessed 26 May 2023.

Punch, vol. 15 (London, 1848).

Punter, David, "Gothic, Theory, Dream," in Charles Crow, ed., *A Companion to American Gothic* (Oxford: Blackwell, 2014), 16–27.

Pushkin, Alexander, *Poems by Alexander Pushkin* [1819]. Trans. Ivan Panin (Boston: Cupples and Hurd, 1888).

Pushkin, Alexander, *The Bronze Horsemen. Selected Poems by Alexander Pushkin* [1837]. Trans. and intro. D. M. Thomas (New York: The Viking Press, 1982).

Puteri Dyung, dir. Abdul Ghafar-Bahari (Kula Lumpur: Penerbitan Mediaseni, 1985).

Radford, Benjamin and Joe Nickell, *Lake Monster Mysteries: Investigating the World's Most Elusive Creatures* (Kentucky: University Press of Kentucky, 2006).

Revenge of the Creature, dir. Jack Arnold (Los Angeles: Universal Pictures, 1955).

Rhee, Suk Koo, "The Erotic-Grotesque Versus Female Agency in Colonial Korea in Park Chan-wook's *The Handmaiden*," in *Canadian Journal of Film Studies/ Revue Canadienne d'études cinématographiques* 29/2 (2020), 115–38.

Rhys-Evans, Peter H. *The Waterside Ape: An Alternative Account of Human Evolution* (Boca Raton: CRC Press, 2019).

Rigby, Kate, "Writing in the Anthropocene: Idle Chatter or Ecoprophetic Witness?," *Australian Humanities Review* 47 (2009), 173–87.

Riviere, Joan, "Womanliness as a Masquerade," in Athol Hughes, ed., *The Inner World and Joan Riviere. Collected Papers: 1920–1958* (London: Karnac Books, 1991), 90–101.

Roberts, Jeremy, *Chinese Mythology A-Z* (New York: London House, 2009).

Romano, Stefania, Vincenzo Esposito, Claudio Fonda, Anna Russo and Roberto Grassi, "Beyond the Myth: The Mermaid Syndrome from Homerus to Andersen. A Tribute to Hans Christian Andersen's Bicentennial of Birth," *European Journal of Radiology* 58/2 (2006), 252–9.

Romney, Jonathan, "*Evolution* Director Lucile Hadžihalilović: 'The Starfish Was the One Worry,'" in *The Guardian*, 28 April 2016. <https://www.theguardian.com/film/2016/apr/28/evolution-lucile-hadzihalilovic-starfish-worry-boys-mothers>. Accessed 24 June 2022.

Rose, Steve, "How Post-horror Movies Are Taking over Cinema," *The Guardian*, 6 July 2017. <https://www.theguardian.com/film/2017/jul/06/post-horror-films-scary-movies-ghost-story-it-comes-at-night>. Accessed 29 May 2021.

Ross, Deborah, "Miyazaki's Little Mermaid: A Goldfish Out of Water," in *Journal of Film and Video* 66/3 (Fall 2014), 18–30.

Rossi, Rikka, "Primitivism and Spiritual Emotions: F. E. Sillanp's Rural Decadence," in Pirjo Lyytikäinen, Riikka Rossi, Viola Parente-Čapkov and Mirjam Hinrikus, eds, *Nordic Literature of Decadence* (New York: Routledge, 2019), 119–36.

Rozwadowski, Helen M., *Vast Expanses: A History of the Oceans* (London: Reaktion, 2018).

Rozwadowski, Helen M., *Fathoming the Ocean: The Discovery and Exploration of the Deep Sea* (Cambridge, MA: The Belknap Press, 2005).

Rush, Eddy, "Bidong Island, Kuala Terengganu, Malaysia," *Malaysian Nomad*, 18 October 2017. <https://www.malaysian-nomad.asia/2017/10/bidong-island-kuala-terengganu.html>. Accessed 6 June 2022.

Russell, Danielle, "Transforming Borders: Resistant Liminality in *Beloved, Song of Solomon* and *Paradise*," in Jessica Elbert Decker and Dylan Winchock, eds, *Borderlands and Liminal Subjects: Transgressing the Limits in Philosophy and Literature* (Cham: Springer, 2017), 105–22.

Sacconi, A. "La tavoletta di Pilo Tn 316: Un registrazione di carattere eccezionale?," in J. Killen, J. Melena and J. P. Olivier, eds, *Studies in Mycenaean and Classical Greek presented to John Chadwick* (Salamanca: Ediciones Universidad de Salamanca 1987), 551–7.

Sastri, Satyavrat, *Discovery of Sanskrit Treasures: Epics and Puranas* (New Delhi: Yash Publications, 2006).

Savoy, Eric, "The Rise of American Gothic," in Jerrold Hogle, ed., *The Cambridge Companion to Gothic Fiction* (Cambridge: University of Cambridge Press, 2002), 167–88.

Sax, Boria, "The Mermaid and Her Sisters: From Archaic Goddess to Consumer Society," in *ISLE* 7/2 (2000), 43–54.

Scales [*Sayidat al-Bahr*], dir. Shahad Ameen (Abu Dhabi: Image Nation, 2019).

Schalk, Sami, "Reevaluating the Supercrip," in *Journal of Literary & Cultural Disability Studies* 10/1 (2016), 71–86.

Schwabe, Claudia, "The Fairy Tale and Its Uses in Contemporary New Media and Popular Culture Introduction," in *Humanities* 5/4 (2016), 81.

Schweid, Richard, *Octopus* (London: Reaktion Books, 2014).

Scribner, Vaughn, *Merpeople: A Human History* (London: Reaktion Books, 2020).

Secret of Kells, dir. Tomm Moore (Kilkenny: Cartoon Saloon, 2009).

Secret of Roan Inish, dir. John Sayles (London: Jones Entertainment Group, 1994).

Sederholm, Karl H. and Jeffrey Andrew Weinstock, "Introduction: Lovecraft Rising," in Karl H. Sederholm and Jeffrey Andrew Weinstock, eds, *The Age of Lovecraft* (Minneapolis: University of Minnesota Press, 2016), 1–42.

Sekien, Toriyama, *Konjaku Hyakki Shūi* [Supplement to The Hundred Demons from the Present and the Past] [c.1781], in *Toriyama Sekien Gazu Hyakki Yagyō Zen Gashū* (Tokyo: Kadokawa Shoten Publishing, 2005).

Shape of Water, The, dir. Guillermo del Toro (Los Angeles, Fox Searchlight, 2017).

Sharf, Zack, "'The Shape of Water' Has Inspired Its Very Own Sex Toy, and It's Selling Out Online," *Indiewire*, 8 February 2018. <https://www.indiewire.com/2018/02/shape-of-water-dildo-sex-toy-selling-out-guillermo-del-toro-1201926708/>. Accessed 28 April 2022.

Showerman, Grant, *The Great Mother of the Gods* (Madison: University of Wisconsin, 1901).

Siew Teip, Looi, Adela Baer and Jali Mohamad, *Temuan: World of words; a Temuan/Malay/English Word List* (Subang Jaya: Center for Orange Asli Concerns, 2016).

Silver, Carole G., "Animal Brides and Grooms: Marriage of Person to Animal Motif B600, and Animal Paramour, Motif B610," in Jane Garry and Hasan El-Shamy, eds, *Archetypes and Motifs in Folklore and Literature* (New York: M. E. Sharpe, 2005), 93–102.

Sins Invalid, "Mission & Vision," *Sins Invalid: An Unshamed Claim to Beauty in the Face of Invisibility*, n.d. <https://www.sinsinvalid.org/about-us>. Accessed 15 May 2022.

Skotak, Robert, "Queen of the Monster Makers: Milicent's Monstrous Masterpieces," in *Famous Monsters of Filmland*, 145/June (New York: Warren Publishing, 1978), 16–21.

Slatton, C. Brittany and Kamesha Spates, *Hyper Sexual, Hyper Masculine? Gender, Race and Sexuality in the Identities of Contemporary Black Men* (London and New York: Routledge, 2014).

Smith, William John, *The Behavior of Communicating: An Ethological Approach* (Cambridge: Harvard University Press, 2009).

Soini, Wayne, *Gloucester's Sea Serpent* (Charleston, SC: The History Press, 2010).

Soles, Carter and Christy Tidwell, *Fear and Nature: Ecohorror Studies in the Anthropocene* (Pennsylvania: Penn State University Press, 2021).

Solomon, Charles. "Foreword," in *Song of the Sea Artbook*, by Tomm Moore (Kilkenny: Cartoon Saloon, 2016).

Somov, Orest, "Rusalka," in Catherine Etteridge, ed., *The Witches of Kyiv and Other Gothic Tales by Orest Somov* [1829] (Lidcombe: Sova Books, 1829).

Song of the Sea, dir. Tomm Moore (Kilkenny: Cartoon Saloon, 2014).

Sonneveld, Reinier, "Incarnations of Death: Leviathan in the Movies," in Koert van Bekkum, et al., eds, *Playing with Leviathan: Interpretation and Reception of Monsters from the Biblical World* (Leiden: Brill, 2017), 280–95.

Sorenson, John, ed., *Critical Animal Studies: Thinking the Unthinkable* (Toronto: Canadian Scholars' Press, 2014).

Spencer, Leland G., "Preforming Transgender Identity in *The Little Mermaid:* From Andersen to Disney," *Communication Studies* 65/1 (2014), 112–27.

Spirited Away, dir. Hayao Miyazaki (Burbank: Walt Disney Home Entertainment, 2001).

Spirited Away, *IMDb*. <https://www.imdb.com/title/tt0245429/>. Accessed 17 June 2022.

Spiropoulou, Angeliki, *Virgina, Woolf, Modernity and History: Constellations with Walter Benjamin* (New York: Palgrave Macmillan, 2010).

Spring, dir. Justin Benson & Aaron Moorhead (Los Angeles: XYZ Films, 2014).

Stasov, V. V., *Ivan' Nikoiaevič' Kramskoj* (Sankt-Peterburg: Tippografija A. S. Suvorina, 1887).

Štejner, Evgenij, "Rusalki Kramskogo i poètika smerti v russkom iskusstve vtoroj poloviny 19 veka," in V. L. Rabinovič and M. S. Uvarov, eds, *Memento Vivere ili Pomni o smerti* (Moskva: Academia, 2006), 168–95.

Stephens, Elizabeth, "Twenty-First Century Freak Show: Recent Transformations in the Exhibition of Non-Normative Bodies," in *Disability Studies Quarterly* 25/3 (2005), n.p.

Stilz, Gerhard, "Australia – the Space that Is Not One: A Literary Approximation," in Gerd Dose, ed., *Australia: Making Space Meaningful* (Tübingen: Stauffenburg, 2006), 29–46.

Strong, Herbert A., *The Syrian Goddess: Being a Translation of Lucian's "De Dea Syria,"* in *With a Life of Lucian* [1913] (London: Forgotten Books, 2015).

Sullivan, Bruce M. and Patricia Wong Hall, "The Whale Avatar of the Hindoos in Melville's Moby Dick," *Literature and Theology* 15/4 (2001), 358–72.

Suvorin, A. S., *Poddelka Rusalki Puškina. Sbornik statej i zametok* (Sankt-Peterburg: Izdanie A. S. Suvorina, 1900).

Szostack, Phil, *The Art of Star Wars: The Mandalorian*, (New York: Abrams Books, 2020).

Takahashi, Rumiko, *Ningyo Shirīzu* [Mermaid Saga] (Tokyo: Shogakukan, 1984–94).

Tatar, Maria, "Introduction: Hans Christian Andersen," in Maria Tatar, ed., *The Classic Fairy Tales* (New York: W. W. Norton, 1999), 212–16.

Taylor, Brian and Michael Zimmerman, "Deep Ecology," in Bron Taylor, ed., *Encyclopedia of Religion and Nature* (London and New York: Continuum, 2008), 456–9.

Taylor, Bron, *Dark Green Religion: Nature Spirituality and the Planetary Future* (Oakland: University of California Press, 2010).

Templesmith, Ben, *The Squidder* (San Diego: IDW Publishing, 2015).

Teodorski, Marko, "Siren Literature, 1870–1920," in Bojan Jović and Tijana Tropin, eds, *Marginalni i marginalizovani žanrovi u književnosti* (Beograd: Institut za književnost i umetnost, 2022).

Thomson, David, *The People of the Sea* (Edinburgh: Canongate Books, 2017).

Times, The, 10 October 1848(a).

Times, The, 14 November 1848(e).

Times, The, 14 October 1848(b).

Times, The, 2 November 1848(d).

Times, The, 21 November 1848(f).

Times, The, 23 October 1848(c).

Toom, Karel van der et al., *Dictionary of Deities and Demons in the Bible* (Leiden: Brill, 1999).

Towrie, Sigurd, "Selkie and Fin: One and the Same?," Orkneyjar, 2021. <http://www.orkneyjar.com/folklore/selkiefolk/origins/origin2.htm>. Accessed 3 May 2021.

Tucker-Jones, Anthony, *Armoured Warfare in the Korean War: Rare Photographs from the Wartime Archives* (Barnsley, South Yorkshire: Pen and Sword Books, 2012).

Tulip, James, "David Malouf as Humane Allegorist," in *Southerly* 41/4 (1981), 392–401.

Tungol, J. R., "Eric Ducharme: Meet Real-Life Man Mermaid, AKA 'Merman,' and Owner of The Mertailor [VIDEO]," in *International Business Times*, 4 February 2013. <http://www.ibtimes.com/eric-ducharme-meet-real-life-man-mermaid-aka-merman-owner-mertailor-video-1171525>. Accessed 13 June 2022.

Turgenev, Ivan Sergeyevich, *A Sportsman's Sketches* [1852]. Trans. Constance Garnett (Scotts Valley: CreateSpace Independent Publishing Platform, 2014).

Ue, Tom, "Narrative, Time and Memory in Studio Ghibli Films," in Bruce Babington and Noel Brown, eds, *Family Films in Global Cinema: The World Beyond Disney* (London: I. B. Tauris, 2014), 223–38.

Underwater, dir. Eubank, William (Century City: 20th Century Fox, 2020).

Untamed, The (*La región salvaje*), dir. Amat Escalante (Mexico: Mantarraya Productions & Tres Tunas, 2016).

Valkenier, Elizabeth, *Russian Realist Art. The State and Society: The Peredvizhniki and Their Tradition* (Ann Arbor: Ardis, 1977).

Van de Kamp, Henk, "Leviathan and the Monsters in Revelation," in Koert van Bekkum, et al., eds, *Playing with Leviathan: Interpretation and Reception of Monsters from the Biblical World* (Leiden: Brill, 2017), 167–75.

Verba, Natal'ja Ivanovna, "Sjužet o morskoj deve v kul'ture XIX veka," in *Muzykal'naja kul'tura v teoretičeskom i prikladnom izmerenii* 4 (2017), 47–53.

Verba, Natal'ja Ivanovna, *Arhetipičeskij obraz morskoj devy v muzykal'noj kul'ture. Tom 1. Dissertacija* (Sankt-Peterburg: Rossijskij gosudarstvennyj pedagogičeskij universitet im. A. I. Gercena, 2021).

Verevis, Constantine E., "Bizarre Love Triangle: The Creature Trilogy," in Claire Perkins and Constantine Verevis, eds, *Film Trilogies* (London: Palgrave Macmillan, 2012), 68–87.

Verne, Jules, *Twenty Thousand Leagues Under the Seas*. Trans. and ed. William Butcher (Oxford: Oxford University Press, 1998).

Vinogradova, L. N., "Rusalka," in C. M. Tolstaja, ed., *Slavjanskaja mifologija. Ènciklopedičeskij slovar', izdanie 2-e ispravlennoe i dopolnennoe* (Moskva: Meždunarodnye otnošenija, 2002).

Visit Faroe Islands, "Kópakonan (Seal Woman)," *Visit Faroe Islands*, 2016. <https://www.visitfaroeislands.com/about/myths-legends/kopakonan-the-seal-woman/>. Accessed 3 May 2021.

Vivarelli, Nick, "Saudi Director Shahad Ameen on Subverting Patriarchal Power in 'Scales'," *Variety*, 3 September 2019. <https://variety.com/2019/film/festivals/venice-film-festival-saudi-arabia-shahad-ameen-scales-trailer-1203321491/>. Accessed 12 February 2021.

Wachtel, Michael, *A Commentary to Pushkin's Lyric Poetry, 1826–1836* (Wisconsin: The University of Winsconsin Press, 2011).

Warner, Marina, *From the Beast to the Blonde: On Fairy Tales and Their Tellers* (London: Vintage, 1994).

Waugh, Arthur, "The Folklore of the Merfolk," in *Folklore* 71/2 (1960), 73–84.

Weisinger, Mort and Paul Norris, "The Submarine Strikes," in Mort Weisinger, ed., *More Fun Comics #73* (New York: Detective Comics, 1941), 30–8.

Whitehead, Dan, *Tooth and Claw: A Field Guide to "Nature Run Amok" Horror Movies* (Manchester: The Zebra Partnership, 2012).

Wiederholt, Emmaly, "Disabled Bodies, Disabled Ways of Thinking: An Interview with Hanna Cormick," *Stance on Dance*, 17 May 2018. <http://stanceondance.com/2018/05/17/disabled-bodies-disabled-ways-of-thinking/>. Accessed 23 September 2021.

Wilhelm Vollmer, *Wörterbuch der Mythologie* [Complete Dictionary of All Peoples] (Stuttgart: Krais and Hoffman, 1874).

Wolfwalkers, dir. Tomm Moore (Kilkenny: Cartoon Saloon, 2020).

Wonderlay, Anthony, *At the Font of the Marvelous: Exploring Oral Narrative and Mythic Imagery of the Iroquois and their Neighbors* (New York: Syracuse University Press, 2009).

Wood, Juliette, "Mermaid/Merman," in Jeffrey Andrew Weinstock, ed., *The Ashgate Encyclopedia of Literary and Cinematic Monsters* (Farnham: Ashgate, 2014), 411–15.

Wood, Robin, "An Introduction to the American Horror Film," in Barry Keith Grant, ed., *Robin Wood on the Horror Film: Collected Essays and Reviews* (Detroit: Wayne State University Press, 2018), 73–110.

Wright, Judith, *Because I Was Invited* (Melbourne: Oxford University Press, 1975).

Wright, Judith, *Preoccupations in Australian Poetry* (Melbourne: Oxford University Press, 1965).

Yamato, Lori, "Surgical Humanization in H. C. Andersen's 'The Little Mermaid,'" in *Marvels & Tales* 31/2 (2017), 295–312.

Yee Yun, Tan, "She Can Be Your Mermaid for $1600," *Asia One*, 1 September 2013. <https://www.asiaone.com/women/she-can-be-your-mermaid-1600?amp>. Accessed 6 June 2022.

Yoaketsugeru Rū no Uta [*Lu over the Wall*], dir. Masaaki Yuasa (Tokyo: Toho, 2017).

Zaharov, N. V., Val. A. Lukov and Vl. A. Lukov, *Dramaturgija A. S. Puškina problema sceničnosti* (Moskva: Izdatel'stvo Moskovskogo gumanitarnogo universiteta, 2015).

Zečević, Slobodan, "Rusalke i todorovci u narodnom verovanju severoistočne Srbije," in *Glasnik Etnografskog muzeja* 37 (1974), 109–39.

Zelenin, D. K., *Izbrannye trudy. Očerki russkoj mifologii: Umeršie neestestvennoju smert'ju i rusalki* (Moskva: Indrik, [1916] 1995).

Zgurskaja, O. G., "Mifologičeskij i real'nyj mir v drame A.S. Puškina Rusalka," in *Voprosy kognitivnoj lingvistiki* 4 (2014), 85–91.

Zimmerman, Michael E., "The Threat of Ecofacism," in *Social Theory and Practice* 21/2 (1995), 207–38.

Zipes, Jack, *Fairy Tales and the Art of Subversion* (New York: Routledge, 1991).

Notes on Contributors

AMYLOU AHAVA is a PhD student and resident punk rocker at Texas Tech University. Her childhood dream of being a horror host stemmed into a lifetime love of everything scary and now has led to a focus in American horror literature, particularly through the lens of disability and mad studies. By day a film professor to undergrads, by night, a writer and reviewer for *Nightmarish Conjurings* and *Daily Grindhouse.*

LEILA ANANI is a lifelong mythology fan and lover of mermaids. She has First Class BA degrees in English Literature, Ancient History and Philosophy. She is a performance poet with nine published poems including winning The Roald Dahl Foundation Wondercrump Poetry award two years running. She currently resides in Suffolk, UK, where she runs a Post Office, and is the longest-serving member on the committee of her local history society. She can be found on Facebook and Pinterest.

SIMON BACON is a writer and film critic based in Poznań, Poland. He has edited books on various subjects, including *Gothic: A Reader* (2018), *Horror: A Companion* (2019), *Monsters: A Companion* (2021), *Nosferatu in the 21st Century* (2022), *Spoofing the Vampire* (2022), *The Anthropocene and the Undead* (2022), *The Palgrave Handbook of the Vampire* (2023) and *The Palgrave Handbook of the Zombie* (forthcoming). He has also published a series of books on vampires in popular culture: *Becoming Vampire* (2016), *Dracula as Absolute Other* (2019), *Eco-Vampires* (2020), *Vampires from Another World* (2021), *1000 Vampires on Screen* (2023), and is currently working on *100 Draculas on Screen.*

RUTH BARRATT-PEACOCK is an Australian literary studies scholar and musicologist. Her research interests include: Romanticism, Australian poetry, anime, and metal music. Her current projects are on classical music as a spatial practice on page and screen (FSU Jena), the social construction of childhood in metal music studies (University of Huddersfield, Walter Benjamin Fellow), and affect in children's media (with Sophia Staite at The Global Sentimentality Project, FAU Erlangen-Nürnberg). Her book, Concrete Horizons: Romantic Irony in the Poetry of David Malouf and Samuel Wagan Watson (Peter Lang, 2020), represents the most comprehensive study of Malouf's poetic oeuvre to date.

DAISY BUTCHER is a scholar attached to the *Open Graves, Open Minds Project* at the University of Hertfordshire. She is the editor of *Evil Roots: Killer Tales of the Botanical Gothic* (2018) and co-editor of *Crawling Horror: Creeping Tales of the Insect Weird* (2021). She has spoken across Europe at numerous conferences and her research looks at the EcoGothic, Post-colonial and Gendered implications of Plant Horror and Egyptian Gothic with particular emphasis on body horror and feminine archetypes.

OCTAVIA CADE is a New Zealand writer. She has a PhD in science communication and a MSc in biology, which she spent in the intertidal zone, studying seagrasses. Much of her academic work looks at the intersection between science and speculative fiction. Relevant work has appeared in such markets as *Horror Studies, Interdisciplinary Literary Fiction, Supernatural Studies, MOSF Journal of Science Fiction* and more. She is the 2023 Ursula Bethell writer in residence at the University of Canterbury.

BRIGID CHERRY, PhD, is an independent scholar, a retired Research Fellow in Screen Media at St Mary's University, Twickenham. Her research is focused on horror film and TV, and horror and SF fan cultures. Recent publications include work on *Doctor Who* and *Twin Peaks* fan memes, graphic novel adaptations of *Alice in Wonderland*, Gothic monstrosity in the films of Tobe Hooper, and depictions of real-world serial killers in crime drama, as well as the book *Cult Media, Fandom and Textiles* (2018), and an upcoming book on the TV series *Lost* (2021).

JENNIFER K. COX completed her MA in Science Fiction and Fantasy Literature at Florida Atlantic University in 2013, and her English PhD at Idaho State University in 2020. She has contributed chapters to essay collections on killer clowns and the FX series *An American Horror Story* and published articles in *Quidditas* and the *Journal of the Fantastic in the Arts*. She still enjoys researching and writing about carnivalesque elements in literature and popular culture, but works by day as a professional editor for a digital marketing company.

ASTRID CROSLAND has never lived out of sight of the ocean. Raised between the Shetland archipelago and Auckland's North Shore, the sea has been her constant companion. Having completed her Master's thesis, *The Gothic Hand*, in 2018, she is currently working on her PhD, *The Witches' Turn: The Magic of Popular Media*, at Auckland University of Technology. Also working in fragrance design, she would evoke the Little Mermaid with a blend of mimosa, ylang ylang and frankincense.

PHIL FITZSIMMONS is currently an independent researcher and consultant in education and organisational learning. Prior to this he was Head of Education (Alphacrucis University College Sydney, Australia), Assistant Dean -Research (Faculty of Education, Business and Science - Avondale University, Australia), Director of Research (San Roque Research Institute, California) and senior lecturer (University of Wollongong, Australia). His current research interests include Australian gothic literature, popular culture and adolescent spirituality.

GERARD GIBSON is a PhD researcher in Cinematic Arts at Ulster University. He has a background in the visual arts and experience as a designer, illustrator and educator. He completed a Bachelor's degree in Graphic Design at The Art College, Belfast. He studied Film at Queen's University, Belfast, gaining a Master in Arts with Distinction. He is particularly interested in combining theory and practice-as-research to explore aspects of cinema that have been previously overlooked because they had been considered the purview of the technician rather than the scholar. His research explores space, place and the material in cinematic horror and is currently focused on the embodied indeterminacy and meaning-making mechanisms of the monstrous, and their often-overlooked epistemological consequences.

BRANDON R. GRAFIUS is Associate Professor of Biblical Studies at Ecumenical Theological Seminary, Detroit. He has published several academic volumes on the intersection of horror and religion, most recently his handbook on film *The Witch* in the Devil's Advocates series (Liverpool University Press/Auteur Publishing). His book for general audiences *Lurking Under the Surface: Horror, Religion, and the Questions that Haunt Us* is available from Broadleaf Books. He is also co-editing *The Oxford Handbook of Biblical Monsters* with John W. Morehead (forthcoming in 2023).

JON HACKETT is Associate Professor in Film and Communications at St Mary's University, Twickenham. He has recently published *Scary Monsters: Monstrosity, Masculinity and Popular Music* with Dr Mark Duffett, for Bloomsbury Academic. He has co-edited *Beasts of the Deep: Sea Creatures and Popular Culture* (2018) and *Beasts of the Forest: Denizens of the Dark Woods* (2019) with Dr Seán Harrington, for John Libbey.

PHILIP HAYWARD is an adjunct professor at University of Technology Sydney, Australia. He has published extensively on mermaids, including his two recent books *Making a Splash* (2017) and *Scaled for Success* (2018). Personal Research Website: http://www.islandresearchph.com/.

AGNIESZKA KOTWASIŃSKA is an assistant professor at the American Studies Center, the University of Warsaw. She specialises in Gothic and horror studies, gender studies and queer theory, and feminist new materialism(s). Her current research interests centre on embodiment in the so-called low genres, Slavic Horror, death, illness and mourning in horror, and schizoanalysis. She has published articles in *Somatechnics*, *Polish Journal of American Studies*, and *Humanities*, among others. She is currently working on her first monograph exploring horror fiction by American women writers.

KODI MAIER (they/them) is a queer Film Studies PhD at the University of Hull. Their most recent chapter, "The Other Maiden, Mother, Crone(s): Witchcraft, Queer Identity and Political Resistance in LAIKA's *Coraline*," appears in *Coraline: A Closer Look at Studio LAIKA's Stop-Motion Witchcraft* (2021), edited by Mihaela Mihalova. Other publications include

"Kids at Heart?: Exploring the Material Cultures of Adult Fans of All-Ages Animated Shows" (*Journal of Popular Television*, June 2019) and "Camping Outside the Magic Kingdom's Gates: The Power of Femslash in the Disney Fandom" (*Networking Knowledge: Journal of the MeCCSA Postgraduate Network*, 2017).

MATT MELIA is a senior lecturer in film, literature and media at Kingston University, where he specialises in cult film, British cinema and the work of Stanley Kubrick and Ken Russell. He is co-editor of the forthcoming *Anthony Burgess, Stanley Kubrick and A Clockwork Orange* (Palgrave MacMillan); *ReFocus: The Films of Ken Russell*, and is co-editor of *The Jaws Book, New Perspectives on the Classic Summer Blockbuster* (Bloomsbury) as well as the forthcoming *The Jurassic Park Book* (Bloomsbury).

LAUREN MITCHELL is a researcher and direct care practitioner. She is the founder of The Doula Project, the United States' first formalised full-spectrum doula programme and is the author of the book *The Doulas: Radical Care for Pregnant People* (Feminist Press, 2016). Her first academic book project, *Alienating Aesthetics: Performance Art and the Medical Imagination*, contends with the ethics of visual and performance culture, narrative, medical history and the limitations of our current societal definition of empathy. She has recently published in *Configurations, Departures in Critical Qualitative Interventions*, and *The Journal of Clinical and Translational Science*, and has held a faculty position at Texas Christian University. She is currently pursuing a Master's in Social Work at the University of Tennessee.

DEBADITYA MUKHOPADHYAY is Assistant Professor of English at Manikchak College, affiliated with the University of Gourbanga, India. Popular Literature and Films, Myths, Adaptations, and Theatre are his areas of interest. His research articles have been published in the peer-reviewed journals like *Muse India, DUJES*, etc. He has recently contributed chapters to the collections *Parenting Through Pop Culture* (2020), *Excavating Indiana Jones* (2020), *Critical Insights: Life of Pi* (2020), and *Children and Childhood in the Works of Stephen King* (2020).

MARTINE MUSSIES writes about The Cyborg Mermaid for The Centre for Gender and Diversity at Maastricht University. Next to that, she is working on a project around King Alfred of Wessex in fanfiction, with support of Leiden University. Besides her research, Martine is a professional musician and illustrator. She studies neuropsychology at the University of Chicago via e-learning. Her interests include autism, Japan, languages, martial arts, medievalism, music(ology) and science fiction. More @ www.martinemussies.nl.

JIMMY PACKHAM is a Senior Lecturer in English at the University of Birmingham. He has research specialisms in American literature, the nineteenth century, maritime writing, and the gothic. His critical interest in folk horror is longstanding, too, and emerges out of research into the presentation of national identity and nationhood in contemporary British gothic fiction. He is a co-convenor of the Haunted Shores research network, review editor for *Gothic Nature*, and the author of *Gothic Utterance* (UWP, 2021) and numerous articles and chapters on gothic and oceanic writing.

ALISON (ALI) PATTERSON, PhD, is a Senior Lecturer in the Department of English and Film and Media Studies Program at the University of Pittsburgh. Her research interests include history on film, animation, digital media and multimodal composition, and disability studies. Her most recent publication and presentations concern critical making in the Film and Media Studies classroom, and her current research project involves animation aesthetics and ornamentalism.

CATHERINE PUGH completed her PhD at the University of Essex and is now a writer and independent scholar. Primarily writing about horror and science fiction across cinema, television, theatre and videogames, she is particularly fascinated by ideas of monstrosity and mental illness versus literary madness. Her research interests concern disability, mental illness/"madness," metamorphic monsters and horror landscapes. She has contributed to various collections, including *At Home in the Whedonverse: Essays on Domestic Place, Space and Life*; *Politics of Race, Gender and Sexuality in* The *Walking Dead: Essays on the Television Series and Comics*; *Vying for the Iron Throne: Essays on Power, Gender, Death and Performance in HBO's* Game of Thrones as well as online journals including *Studies in Gothic Fiction* and *Aeternum: The Journal of Contemporary Gothic Studies*.

LAURA SEDGWICK is currently studying for a PhD in Film Studies at the University of Stirling, on the topic of "Haunted Spaces in Contemporary Horror Cinema: Set Design and the Gothic." She is the co-author of *Gothic Dissections in Film and Literature: The Body in Parts* (2017), with Ian Conrich. Her research interests include horror cinema, art history, Gothic studies, cemetery architecture, and British folklore. She also writes Gothic horror and dark fantasy fiction, and she is the host of the Fabulous Folklore podcast.

MANAL SHALABY is a Lecturer in English Literature and Film Studies at Ain Shams University in Cairo, Egypt. She worked at Williams College in Massachusetts, USA from 2013 to 2015 as part of the Fulbright FLTA Program and received her PhD in 2018 from Ain Shams University. Her recent research interests include exploring the relationship between Middle-Eastern and European mythology in ancient and modern folklore, and examining trauma discourse in post-Arab Spring fiction and film.

TOM SHAPIRA is a PhD candidate at the Tel-Aviv University School of Cultural Studies. Tom is the author of *Curing the Postmodern Blues: Reading Grant Morrison and Chris Weston's The Filth in the 21st Century* (Sequart, 2013) and *The Lawman* (PanelxPanel, 2020). His research had appeared in collections such as *Utopia and Dystopia in the Age of Trump* (Fairleigh Dickinson University Press, 2019) and *The Ages of the Flash* (McFarland, 2019).

ALISON SPERLING is currently an IPODI Postdoctoral Research Fellow at Technische Universität Berlin and an Affiliate Fellow at the ICI Institute for Cultural Inquiry Berlin. She also the theory tutor in St Joost School of Art and Design's "Ecology Futures" Programme in the Netherlands.

MARKO TEODORSKI is Research Associate at the Institute for Literature and Arts, Belgrade. He graduated classical archaeology from the University of Belgrade (Serbia) and obtained a PhD in cultural studies from the University of Perpignan (France) and University of Tübingen (Germany). His research primarily focuses on psychoanalysis and literature/literary theory, but spreads over a range of topics: monster theories and studies; ancient Greek culture and philosophy, as well as their modern reception; Victorian literature and culture; ancient Hindu yogic philosophy and practice. Within all these fields, he is particularly interested in the notion of monstrosity as a cultural palimpsest and in historical and cultural forms of the transcendental as an extra-discursive experience. He is the author of *Nineteenth-Century Mirrors: Textuality and Transcendence* (2021) and the editor of *Monstrosity from the Inside Out* (2014). He also plays a lot of video games.

TOM UE researches and teaches courses on nineteenth-century British literature, intellectual history, and cultural studies at Dalhousie University. He is the author of *Gissing, Shakespeare, and the Life of Writing* (Edinburgh University Press, forthcoming) and *George Gissing* (Liverpool University Press, forthcoming), and the editor of *George Gissing, The Private Papers of Henry Ryecroft* (Edinburgh University Press, forthcoming). Ue has held the prestigious Frederick Banting Postdoctoral Fellowship, and he is Honorary Research Associate at University College London.

KEVIN WETMORE, JR. is the author of over a dozen books, including *Post-9/11 Horror in American Cinema, Eaters of the Dead: Myths and Realities of Cannibal Monsters*, and *The Theology of Battlestar Galactica*, as well as over a hundred book chapters, journal articles and essays, on topics from ghosts on the Japanese stage to African Adaptation of Greek tragedy to Shakespeare in graphic novels. He is also the twice-Bram Stoker Award-nominated editor of books such as *Uncovering Stranger Things* and *The Streaming of Hill House*. He is an actor, director and fight choreographer who lives and works in Los Angeles.

JUSTIN WIGARD (pronounced "Why-Guard") is a Postdoctoral Research Fellow in the Distant Viewing Lab at University of Richmond, where he works and teaches in the areas of popular culture, game studies, comic studies, children's literature and digital humanities. He is co-editor of *Attack of the New B Movies: Essays on SYFY Original Films* (McFarland Press, 2023). He has published games scholarship in Cinergie, Unbound, Vault of Culture, and various edited collections. You can find more of his work at his website, justinwigard.com.

CARL WILSON is a contributing guest writer for the Eisner-nominated comic book publishers Fanbase Press and a former Film Editor for PopMatters. com. He has published work in over a dozen titles and has chapters forth-coming in the area of transmedia convergence, including the Indiana Jones franchise, the Uncharted video game series, the depowering of DC super-heroes, the representation of women in Batman games and the digital legacy of Superman. Examples of his work can be found at www.carl-wilson.com.

Index

Genre Fiction and Film Companions

Series Editor: Simon Bacon

The *Genre Fiction and Film Companions* provide accessible introductions to key texts within the most popular genres of our time. Written by leading scholars in the field, brief essays on individual texts offer innovative ways of understanding, interpreting and reading the topics in question. Invaluable for students, teachers and fans alike, these surveys offer new insights into the most important literary works, films, music, events and more within genre fiction and film.

We welcome proposals for edited collections on new genres and topics. Please contact baconetti@googlemail.com or oxford@peterlang.com.

Published Volumes

The Gothic
Edited by Simon Bacon

Cli-Fi
Edited by Axel Goodbody and Adeline Johns-Putra

Horror
Edited by Simon Bacon

Sci-Fi
Edited by Jack Fennell

Monsters
Edited by Simon Bacon

Transmedia Cultures
Edited by Simon Bacon

Shirley Jackson
Edited by Kristopher Woofter

Toxic Cultures
Edited by Simon Bacon

Magic
Edited by Katharina Rein

The Undead in the 21st Century
Edited by Simon Bacon

The Deep
Edited by Marko Teodorski and Simon Bacon

www.ingramcontent.com/pod-product-compliance
Lightning Source LLC
Chambersburg PA
CBHW071832270326
41929CB00013B/1970